D1531968

MARKETING

STRATEGIES FOR

THE NEW

ECONOMY

MARKETING STRATEGIES FOR THE NEW ECONOMY

LARS TVEDE and PETER OHNEMUS

JOHN WILEY & SONS, LTD

Chichester · New York · Weinheim · Brisbane · Singapore · Toronto

Other Wiley Editorial Offices

John Wiley & Sons, Inc., 605 Third Avenue,
New York, NY 10158-0012, USA

Wiley-VCH GmbH, Pappelallee 3,
D-69469 Weinheim, Germany

John Wiley & Sons Australia, Ltd, 33 Park Road, Milton,
Queensland 4064, Australia

John Wiley & Sons (Asia) Pte Ltd, 2 Clementi Loop #02-01,
Jin Xing Distripark, Singapore 129809

John Wiley & Sons (Canada) Ltd, 22 Worcester Road,
Rexdale, Ontario M9W 1L1, Canada

British Library Cataloguing in Publication Data

A catalogue record for this book is available from the British Library

ISBN 0-471-49211-6

Typeset in 11/14pt Bembo by Dorwyn Ltd, Rowlands Castle, Hants.
Printed and bound in Great Britain by Bookcraft (Bath) Ltd, Midsomer Norton, Somerset.
This book is printed on acid-free paper responsibly manufactured from sustainable forestry, in
which at least two trees are planted for each one used for paper production.

CONTENTS

INTRODUCTION

It is now quite a few years ago that several newspapers across the world described an interesting study, which showed that households in the developed world typically had 60–120 electromotors. Most people who read those articles were probably a bit surprised: 60–120 electromotors seemed a lot, particularly when you considered that most people have probably never actually walked into a shop and asked for an electromotor. 'Yes Sir, we have a special offer this week. You can get 12 electromotors for the price of 10.' Never happened.

But check it out for yourself, and you will probably find that this (now old) information was credible after all. Wrist watches, hi-fi equipment, alarm clocks, electric toothbrushes, fridges, VCRs, camcorders, etc. They all had electromotors. Most readers of this book probably have more than 200 by now.

Electricity is a way of moving energy. It was utilized initially to provide power to factories, but it didn't take long before smart people thought of taking it into private homes. Many business executives originally thought that only wealthy households would have electricity, and that the power lines would lead into a single, central

x INTRODUCTION

machine in each house, which you could use for different purposes. The Sears Roebuck catalogue featured, for instance, such a central 'home motor' in 1918. That's not what we got, though. The power lines went into the households all right (wealthy or not), but then into every second room, and then every room, and then to several plugs in each room. And we guess that's how we ended up with 60–120 (or 200) electromotors in a household, powered by power lines, solar panels, batteries and a host of other ways.

So why do we find this so interesting? Because we believe the same will happen to the internet and other online networking. The Internet will be available via wire lines, cable, terrestrial wireless, satellite and mobile phone networks. Data will flow as freely as electricity. No: it will flow *more* freely, since we can transmit data through air and space, while our power supply can travel only over physical wires. The internet will be an integral part of our lives whether we get the content in current form (streaming) or stored form (cached or saved).

We believe that one day, perhaps 20 years from now, we will read in the newspaper that every household has 60–120 (or 200) internet connected devices, and then we will show our children an old, dusty book that we wrote for John Wiley & Sons back in 2001. 'See, kids. We knew that we would get Internet anywhere!' (Perhaps 40 years from now it might be our 200 computers that discuss the newspaper while we are hiding from them in the cellar. But that's another story).

The internet is a way of moving data. This book is about how this ever present flow of data to all these connected devices is changing the rules of marketing. We don't assume they will change *all* the rules, of course, and there are probably a few industries where they will have very little impact, if any. But many rules are changing, some in a dramatic way. What we are describing in this book is the new marketing paradigm evolving in a connected world, where branding, intellectual capital and time to market are the keys to success.

Lars Tvede
Peter Ohnemus

Zug, Switzerland

BEGINNINGS

'I think there is a world market for maybe five computers.'

Thomas Watson, Chairman of IBM, 1943

CHAPTER 1

BEGINNINGS

*O*nce upon a time, there lived a famous economist called Joseph Schumpeter. Born in Austria but spending many years in the UK and USA, Schumpeter was an interesting character. He became known for challenging a librarian to a duel with swords (and winning), insisting on always eating dinner in shirt and tails – even as a student – and for his brief career as the finance minister of Austria. Even more, he was known for stating publicly that his aim was to become the 'best lover in Vienna', 'best horseman in Austria' and 'best economist in the world'. While it is unclear whether he succeeded on his first two aims, and few would claim that he succeeded in the third, he is today recognized widely as the leading thinker of innovation, and the roles entrepreneurs play in shaping our economies.

One of Schumpeter's basic observations was that innovation tends to come in waves. The past four centuries have seen one economic revolution after another. There was the Commercial Revolution, where people moved from the countryside to towns, and started using money to trade locally and internationally. Then came the Industrial Revolution – steam, spinning machines and all that. Then the so-called

Second Industrial Revolution of steel and chemicals, and the Third Industrial Revolution of electromotors and internal combustion engines.

Schumpeter described how each of these, and other, revolutions began with a few important core innovations, and how these would then lead to a chain reaction of new discoveries, investment, spending and lending. Most of the innovations would be in new technology, but they could also be commercial innovations (the economist Robert Solov, an economist working at MIT, calculated in 1957 that 80% of all economic growth comes from technology). Canals, steam engines, spinning machines, electricity, telephony, railways and cars were all examples of technical innovations.

The revolutions would each change the way we work. The Danish Institute for Future Studies exemplified this through an investigation of the sectors that Danish workers had worked in, with a forecast towards 2010. It showed significant transitions (Figure 1.1).

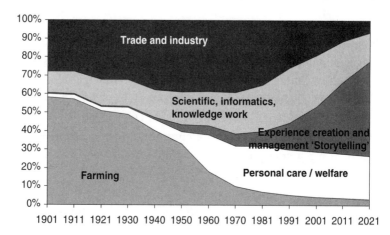

Figure 1.1 Transitions in the Danish workforce allocation, 1901–2021 (forecast). Source: The Copenhagen Institute for Future Studies.

The study showed how the dominance of sectors had changed over time. Almost 60% of the population had worked in agriculture by 1901, but this had been overtaken by industry and trade, and then by service/care taking and informatics. The study predicted that a new role called 'storytelling' would become the biggest employment segment within a few years.

Schumpeter was particularly interested in the purely economic results of innovation. He observed how the waves of innovation and the derived activity would often lead to excesses, bubbles and crashes, sometimes during several financial cycles within the period where the core innovation was rolled out across society. The railway boom led, for instance, to a whole series of stock market crashes each of which was followed by new bull markets.

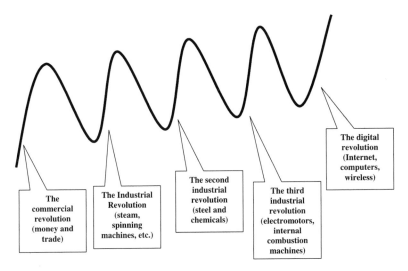

Figure 1.2 The major waves of innovation. The economy is characterized by large waves of innovation that create deviations around the long-term growth curve. Schumpeter called these waves 'Kondratieff', after a Russian economist who first described them.

RIDING THE WAVES

The Internet and computing are jointly creating one of Schumpeter's waves of innovation. By year 2000, less than 5% of the world's population had access to the Web, and most of those used only one or two devices to connect. So there is room for growth. Massive growth, in fact.

We believe that there will be over two billion active online devices in the digital economy by 2008. This will build, change and destroy industries at a speed that no one ever thought possible just a few years ago.

The main cause of innovation is the pursuit of the old scientific principle of reduction. Just think about it: science happened when we learned how to reduce complex problems to their smallest components and then to treat these individually. Break each problem down to its smallest elements, analyse each of these, and then aggregate up to the whole. The commercial and industrial revolutions were built on the same core principles. Commerce took off when man learned to break value down to discrete units, called money. The same thing happened with the Industrial Revolution. Books, for instance, were hand-written by monks before Johann Gutenberg thought of making each letter into a discrete production tool unit and was then able to reduce the production time by combining these units, i.e. Gutenberg used reduction as the step to mass production.

This is also what we do with digital data processing. We reduce everything to bits. Bits are then aggregated into packages, which are integrated into software objects, which are then combined into programs that go into products. The result is thousands, if not millions, of innovations. These innovations shape the digital economy and are the basis for new marketing paradigms.

And new paradigms seem to be needed – strange things have happened since the Internet took off. Things that can't possibly be explained by conventional marketing philosophy. New companies have come out of nowhere with business models based on giving away some or even all of their products. Some of them have then attracted huge amounts of money and seen their stock prices rocket sky high, while most pundits have stared on in amazement. Quite a few of these companies have then gone bust, but others have become profitable – initially just slightly, but then in some cases hugely profitable. They have managed the transition from a garage start-up to the status of global empire within a very short span of years. This has never happened before. It's the beginning of something new.

This is the 'new economy'. An economy that has created fast wealth like never before. It has created economic growth without inflation, with employees making millions on stock options, access to all products at your fingertips, and CEOs in tennis shoes. It's a wonderful thing.

The new economy can also be ruthless though. A rule of thumb among venture capitalists is that you can expect the following outcomes from every 10 000 business plans for high-tech and Internet start-ups:

- 60 of the 10 000 proposed businesses (0.6%) will get funding from venture capitalists.
- 30–40 of those who get funded will go bankrupt or create little value.
- Six of those that succeed commercially will go public – just 0.06%
- One of those that go public will have a star performance after its IPO – 0.001%.

That is one major success out of every 10 000 proposed businesses. It is just one out of 60 companies that received funding in the first place.

Something has changed in marketing. Upsides are bigger. Speed is furious. Teenagers become millionaires. Failure abounds. But not everything has changed. It is probably more correct to say that marketing is yet again going though an evolutionary change, in which we have to add to our knowledge but shouldn't throw everything we learned before out of the window. Marketing as a discipline has never stood still, and it will always evolve and emerge as new commercial phenomena evolve. This is perhaps seen most clearly if we look briefly at how it evolved before the Internet.

THE DISCOVERY OF MARKETING

Marketing has been practised for as long as business has existed, but it didn't become a recognized independent economic discipline until the beginning of the last century – in 1902 to be exact. That year, for the first time ever, students could go to a university and take a course that taught them specifically about marketing. This was at the University of

Michigan, and the term that was used for the subject was 'distributive industries'. 'Marketing' entered the vocabulary when the University of Pennsylvania offered a course entitled 'The Marketing of Products' in 1905; the term became commonplace between 1906 and 1910.

Early marketing theory could, until around 1910, largely be characterized as the period of basic discovery, where the discipline borrowed most of its concepts from general economics. The following decade was more a period of conceptualization, when marketing developed its own vocabulary and ideas and broke free from general economics.

A number of valuable attempts to integrate marketing concepts into a whole appeared around 1920. Then, as now, it was not only universities that were developing marketing theories. Some of the leading marketing models were the brainchild's of commercial marketing companies such as J. Walter Thomson, which was America's largest advertising bureau at the turn of the century. The Walter Thomson agency decided to integrate all existing marketing theories at the time into a managerial tool, which they called 'T-square'. They used this tool to advise their large clients not only about advertising, but also about overall marketing strategies. This inspired other advertising agencies to embrace the new theories, and to develop their own theories, which they used as a basis for their transitions into 'full-service' marketing agencies. Some of the first major clients to be inspired by these new full-service agencies were Pennsylvania Railroad, Corning Glass Works, General Foods and Kodak.

Stanley Latshaw is often described as the man who inspired the beginnings of modern marketing research. Latshaw was an advertising salesman with the Curtis Publishing Company, but he was dissatisfied with the way ad space was sold. The problem was that the advertising buyers did not have access to information about the audience for the publications where they advertised. Determined to resolve the problem, he managed to convince his boss, Mr Curtis, that it would be profitable to publish studies of the market for their publications so that they could attract more advertising. They hired Charles Coolidge Parlin, a schoolmaster, to do the job. Parlin made a number of large market research studies and managed, among other things, to create the foundation for increased advertising revenues in the *Saturday Evening Post*. In 1916, the United States Rubber Company established a market research department, and other companies followed suit.

THE TRADITIONAL MARKETING SCHOOLS

The knowledge in any philosophical or academic discipline is normally divided into so-called 'schools of thought'. A new school of thought is not identified as such on the very day when the first theory within it emerges. You can't talk about a forest when there is only one tree. But as more and more credible theories are launched, it becomes clear that they belong together in a cluster, and after a while they are given a name.

One of the first leading marketing schools of thought was the *Commodity School*. This emerged around 1910 and gained broad acceptance just after the Second World War. This school was focused on the physical characteristics of products and services, and on how customers responded to these characteristics. The main disciplines were:

- field research
- product/service positioning
- pricing strategies
- distribution strategies
- promotion strategies.

Another approach, the *Functional School*, which classified the different tasks involved in the marketing practice, dealt with questions such as: 'What are the main tactical tasks to be performed for international marketing of product X, and how should you organize their execution?' The tasks included:

- basics of marketing information retrieval
- organizing the information flow
- sector evaluation
- competition strategies
- product/service positioning

- market positioning
- pricing strategies
- writing marketing plans
- the marketing planning cycle.

The Functional School, which emerged around the same time as the Commodity School, has evolved little on a theoretical basis since the 1950s, but it has been influential in terms of clarifying marketing roles and planning tasks.

The *Regional School*, which developed after 1913, dealt with distribution issues – how to bridge the gap between buyers and sellers. It would, for instance, address the question of how to optimize location and size of wholesale and retail outlets for a company's products. It was relevant for all distribution strategies.

The *Institutional School* focused on the organizational implications of marketing – how to organize a company internally, and how to cooperate with other companies to reach success. The school emerged in the 1910s, largely as a response to the consumers' lack of understanding of the gaps between farmers' sales prices and retail prices for agricultural products. Its concepts clarified how a corporation or institution should organize itself for marketing success. An example of early adoption is Procter & Gamble, which was probably the first company to designate the title 'Marketing Manager' to an employee. The school's disciplines included:

- basics of marketing information retrieval
- organizing the information flow
- the purpose of the marketing audit
- the marketing planning cycle.

The terms 'marketing management' and 'marketing strategy' were probably first introduced by Leverett S. Lyon in 1926; this concept developed eventually into the *Managerial School*, which became increasingly important from the late 1940s. This school, which was a part of the managerial economics movement that took off in the late 1940s and early 1950s, provided studies of how a manager could change concepts into tasks and manage these tasks. The key concepts addressed were:

- basics of marketing information retrieval
- organizing the information flow
- the purpose of the marketing audit
- defining strategic company objectives
- business environment reviews
- sector audits
- competition audits
- product/service positioning audits
- pricing strategy audits
- distribution strategy audits
- promotion strategy audits
- writing marketing plans.

The last of the six main marketing schools was the *Buyer Behavior School*, which focused on the reasons why people would buy certain products and services. It addresses questions such as: 'What are the exact reasons that customers in segment X prefer to buy product Y?' This was particularly important for the development of marketing research.

THE EMERGENCE OF A NEW SCHOOL

Most of the mainstream knowledge gained from these six marketing schools is still true and relevant for the theory and praxis of marketing, whether it is within the 'old' or 'new' economies. Most of it will, in fact, probably be relevant for as long as business exists. But there are phenomena in the economy that these schools don't cover because the phenomena didn't exist or weren't important when the schools emerged. These new phenomena arise when companies, people and computers are tied together in huge digital networks where time, place and legacy investments matter less and less. None of the previous marketing schools had assumed the effects of millions of computers, billions of small chips, armadas of communication satellites – and the Internet. But while the traditional marketing schools did not describe the new networking phenomena, there have subsequently been numerous attempts to encapsulate many of the changes that have oc-curred. Enormous amounts of literature have already been written about the effects of digital networking, and anyone who reads some of it will soon see that it describes things that are truly new and different. It is beginning to look like a new school of marketing.

We could call this new school many things, but the best term is perhaps the 'Digital School' (Figure 1.3). Networking has always been a part of marketing, so this term doesn't encapsulate the school. It's the *digital* aspect that is new. Digital communication is responsible for a cluster of new phenomena that are so strong and different that we think they deserve their own school.

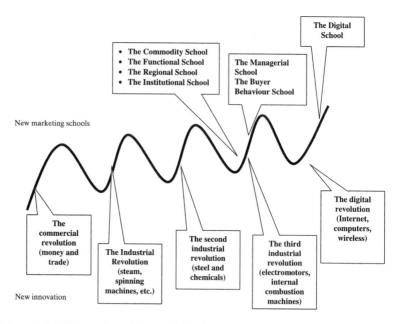

Figure 1.3 Waves of innovation and the development of marketing schools.
The first six marketing schools described the effects of a commercial and industrialized society. The Digital School describes aspects that occur as a consequence of computing and digital networking.

This Digital School of Marketing has received contributions from numerous great thinkers. We shall meet many of them as we proceed further in the book, but it seems fair to list a few right here:

- *George Gilder* wrote widely on numerous phenomena in the digital economy, and originated 'Gilder's Law', which states that the total bandwidth of telecommunication triples every 12 months.
- *James Moore* coined the term 'business ecosystems'.
- *Robert Metcalf* was the originator of 'Metcalf's Law', which stipulates that the value of a network grows in exponential proportion to the number of people connected.
- *Gordon Moore* was the originator of 'Moore's Law', which suggested that chips' processing powers would continue to grow exponentially for a sustained period of time.
- *Peter Drucker* argued that in the new economy, the key means of production was no longer capital equipment but human intellect. This shifted the power from corporations to individuals. Drucker was also one of the first to describe how new technologies in the digital economy typically have to be at least 10 times better than what they replace in order to succeed.
- *Brian Arthur* published a series of influential articles about network effects, increasing returns, feedback processes and path dependency. 'Brian Arthur's Law' states that companies operating in the network economy often experience increasing marginal returns.
- *Stan Davis and Bill Davidson* were among the first to describe the concept of a market segment of one.
- *Geoffrey Moore* provided the best description of the critical development phases in a digital economy and coined the now widely used economic expressions 'crossing the chasm', 'gorilla game' and 'tornado'.
- *Alvin Toffler* coined the term 'prosumer' to describe the consumer who contributes to the production process through collaboration with the supplier.
- *Mark Weiser* was first to write about 'ubiquitous computing'.
- *Paul Romer* developed a number of simple theories and models that described the digital economy in academic terms.
- *Paul Krugman* made numerous observations of the digital economy, and stressed that supply and demand curves often had the opposite slopes in the new economy compared with the old economy.

- *Paul Saffo* suggested that we talk about the 'value web' instead of the 'value chain', since value flows both ways in the digital economy. Saffo also coined the term 'disinterremediation' to describe the fact that digital networking changes intermediation rather than eliminating it.
- *Alvert Bressand* coined the term 'R-Tech' for technology that facilitates relationship management (creating tailored products, recalling preferences, anticipating interests).
- *John Hagel* suggested that technology should not only help customers generate information about users, but also help users generate information about themselves and about economies.
- *James Gleich* wrote the book *Faster: the acceleration of just about everything*, in which he described how businesses (and other things) are accelerating frantically.

What these and many other writers have described is a new economy. But what *is* this new economy? What are the drivers behind the changes that we are experiencing? And how do the new phenomena change the challenges we meet as marketers?

The first step towards understanding these questions is to look at those things that are truly different now to what they were, say, 30 years ago. People don't change, but our inventions do. So let's start our endeavours with a look at the new technologies that seem to drive it all.

TECHNOLOGY

'Computers are incredibly fast, accurate, and stupid; humans are
incredibly slow, inaccurate, and brilliant; together they are powerful
beyond imagination.'

Albert Einstein

CHAPTER 2

TECHNOLOGY

*T*he phenomenon that we call the new (or digital) economy began in small pockets of our world, and has since grown relentlessly to the point where it is now difficult to think of any industry that isn't somehow affected by it. Nor is it possible to read about business and economics without coming across numerous articles about huge companies thriving on the new economy – companies that were not even around a few years ago. Most of these masters of the new universe come from one of three different sectors – technology, media and telecommunications. These are sometimes called 'TMT companies' or 'TIME companies' (telecommunications, Internet, multimedia and entertainment); they are the companies that, more than any other, have demonstrated how different the digital economy is. So where did these TMT companies come from? Three occurrences worked together to create the basis for a revolution:

- The first change was a wave of *core innovation* in microprocessors, software and telecommunication systems. Core innovation is the hard work that is done at universities, in large corporations, and in start-ups, and which fundamentally changes the way things are done.

- The second change was the increased use of *open standards*. Pursuit of open standards is basically a business model, whereby you interface your technology to common layers that many players agree on. This makes it easier to make different devices and services work together smoothly.

- The third change was *deregulation* within the telecommunications and media sectors. Deregulation is a government approach to the economy that aims to stimulate competition and cut red tape.

Each of these fundamental drivers stimulated interesting new developments (Figure 2.1):

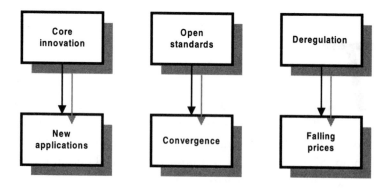

Figure 2.1 Fundamental changes leading to the digital revolution. It was primarily an explosion of new core innovation, a move towards open standards and a wave of deregulation that created the conditions necessary for the new economy.

- Core innovation in technology made it possible to create a multitude of new and compelling *applications*. No one thinks they need a chip, for instance, but they do like to use the telephone, PC, car and TV – all of which depend on chips.

- Open standards meant *convergence*, meaning that many companies developed their technology so that it was compatible. They developed open application programming interfaces (APIs), which enabled different suppliers to become mutually compatible, e.g. mobile phone users could connect with each other, even if the phones were different brands (actually, this is not always the case in the USA, but never mind).

- Deregulation meant the end of monopolies. And what happens when monopolies face competition? They reduce prices to retain market share, and you get *falling prices*.

The three initial drivers gave rise to lots of cool new tools and toys (applications), which worked together (convergence) and were getting pretty cheap (falling prices). What did that give? It created *user communities*, which meant growing streams of *revenues* (Figure 2.2).

Figure 2.2 illustrates what seems to be a fairly simple set of relationships, where three great initial changes led to the development of new markets. This does not seem terribly different from what had

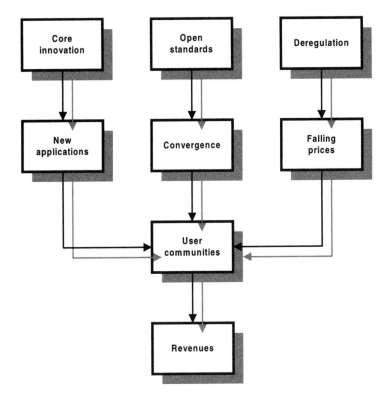

Figure 2.2 User communities. New applications, convergence and falling prices attracted more and more customers who began interacting through user communities.

happened many times previously. But it is. The digital economy is driven by phenomena that are vastly different from anything previously experienced. Never before has mankind experienced markets where prices fall so quickly, product cycles are so short, product performance grows so quickly, borders break down so quickly, people connect so freely, and tiny start-ups develop into global empires within a few years. Something is definitely different now.

So let's try to take a closer look. We can start by studying the process of core innovations and how the roll-out of new applications in the new economy evolves.

CORE INNOVATION AND NEW APPLICATIONS

The combination of numerous core innovations and new applications provides one of the three driving forces behind the development of the digital economy (Figure 2.3). A key element here is a tiny flake of melted sand – silicon. This virtually free raw material can be used to create the most fantastic products mankind has ever conceived.

Man-made Brains

Silicon is the main raw material of the central processing unit of computers, 'chips'. The electric switches in chips are called transistors. Think of chips as brains and the transistors as brain cells. Chips process data. This processing is measured in millions of instructions per second, or MIPS. One of the key factors in increasing the number of MIPS is to make each chip as small as possible. The first computers didn't use chips, and they weighed several tons. Each transistor was a large object that you could hold in your hand to inspect it. Today you need a microscope if you want to see what a transistor looks like. You will soon need a microscope to see what most chips look like.

It is probably fair to say that the king of chips was a man named Gordon Moore. He was a cofounder of Intel. Moore noticed one day in 1965 that historically the capacity of microchips had doubled each

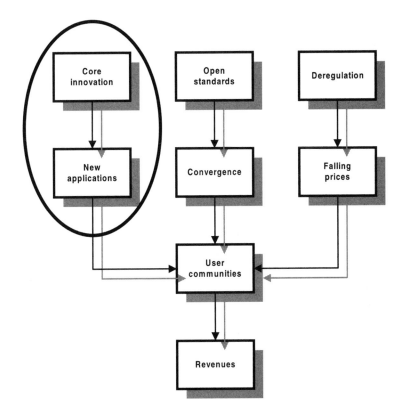

Figure 2.3   Core innovation and applications in the digital economy. The development of core innovation and applications in the digital economy follows rules that are truly different from those we have seen before.

year. He described the phenomenon in an article for the 35th anniversary issue of the magazine *Electronics* (19 April 1965). The article, 'Cramming more components onto integrated circuits', claimed that:

> The complexity for minimum component costs has increased at a rate of roughly a factor of two per year. Certainly over the short term this rate can be expected to continue, if not to increase. Over the longer term, the rate of increase is a bit more uncertain, although there is no reason to believe it will not remain nearly constant for at least 10 years. That means by 1975, the number of components per integrated circuit for minimum cost will be 65,000. I believe that such a large circuit can be built on a single wafer.

The pace of change has slowed down a bit since, so the rule has been modified (with Gordon Moore's approval) to reflect that the doubling occurs only every 18 months. So Moore's Law now states that: 'The pace of microchip technology change is such that the amount of data storage that a microchip can hold doubles every year or at least every 18 months.' Although the rule is slightly more conservative than described in 1965, the doubling of capacity every 18 months is of course still very powerful indeed. So, it is no wonder that we begin to consider when computers will become truly intelligent. Perhaps not intelligent in the same way as we are, but at least with a data-processing capacity equal to that of our own brains.

A good way to grasp how smart the chips really are is to consider how well they cam mimic something that people do all the time: visual recognition. Experience shows that a computer with a capacity of one MIPS can identify very simple patterns in a real landscape. It might, for instance, be able to spot straight lines like coloured sticks in a forest. With 10 MIPS, it becomes able to spot grey-scale patterns. This is how a cruise missile can home in on its target. Move to 100 MIPS, and the computer can enable a car to follow a road by itself. Ten thousand MIPS, and it can identify three-dimensional objects in a 3-dimensional landscape.

IBM's 'Big Blue' computer, which defeated chess master Kasparov in 1997, used approximately 3000 MIPS for the task. A lizard handles approximately 1000 MIPS, about the same as the fastest PCs in 2001. A monkey's brain is somewhere around 1 000 000–10 000 000 MIPS, with a processing power equivalent to 1000–10 000 of the fastest current PCs combined. A human is closer to 100 000 000 MIPS. Kasparov couldn't defeat the inferior IBM machine in 1997 because the human brain cannot allocate more than a tiny fraction of its processing power for a single task, such as chess. Brains are structured very differently from chips. Brains are intuitive, while computers are single-minded.

The largest computers of 2001 are already on a par with a monkey's brain. Fortunately, although these computers do many things, such as playing against world-class chess players, they have not yet shown any signs of consciousness. But they can move around like robots and do some of the things that humans do. One of the many books written about the subject is *Robot* (1999) by Hans Movarec, who leads the robot research programme at Carnegie Mellon University. Movarec predicts that the most sophisticated computers will reach the processing power of human brains by around 2040, which sounds exciting. By 2050, he thinks they will be far ahead of us (which sounds a bit scary). Movarec points out that computer design and production is itself based on the use of computers, and although humans are in the loop, we long ago entered into a development process in which computers develop computers, and in which they will increasingly learn to feed and repair themselves (the robots in his institute walk around in the corridors; once they run low on battery, they look for a plug in the wall, go to it, and recharge themselves).

Chips will continue to improve at a breathtaking speed. Much of this improvement will take place through numerous developments within the existing core concepts, but there are also huge leaps under way – quantum computers, opto-electronics, and nano technology are just a few examples that await us.

Quantum computing is a bit complicated to grasp and explain (to put it mildly), so we will skip that. Opto-electronics is a technology that uses light waves instead of electromagnetic waves to transmit signals. This works well on a telephone line today, but it is more difficult (but not impossible) to zoom it into a chip. Nano technology will probably be needed when we want to adapt opto-technology. Nano technology was launched in a now famous speech by physicist and Nobel Prize winner, Richard Feynman, on 29 December 1959, at the annual meeting of the American Physical Society at the California Institute of Technology (Caltech).

'The principles of physics, as far as I can see, do not speak against the possibility of manoeuvering things atom by atom,' Feynman said. 'We need to apply at the molecular scale the concept that has demonstrated its effectiveness at the macroscopic scale: making parts go where we want by putting them where we want!'

The human body is made up of atoms; these atoms have a combined value as commodities of about $2. Any residual commercial value of a person comes from the way the atoms are structured. But if we can develop tiny, microscopic machines to put each and every atom exactly where we want it to be, then we can take miniaturization to the ultimate level. Current technology, it has been said, is like someone building LEGO™ models with boxing gloves on. Nano technology will be like taking the gloves off and actually snapping the blocks together in the ideal pattern.

The concept of nano technology for development of computer chips took a new twist when researchers from Lucent and Bell Labs announced in 1998 that they were working on chips based on modified DNA strings. DNA is the molecule that carries the code for biological organisms, and the Lucent/Bell teams now believe that they are able to manufacture chips that use DNA to achieve chip performance that is about 1000 times better than the best chips of 2001. That pattern could be the world's ultimate computer chip. It seems that Hans Movarec was on to something.

Swarms of Frozen Reflexes

As we said earlier, we can think of chips as brain cells. Brain cells can be part of something very complex, such as the human brain, but they can also comprise something much simpler, for example an ant's brain. Ants outnumber we human beings by a very large majority. This is similar to the situation with computer chips. The simplest of these chips are called 'jellybeans'. These are tiny stores of instructions for action, just like batteries are tiny stores of energy. Jellybeans already outnumber the bigger computer chips by a factor of 30 : 1, and this ratio is growing. Jellybeans are the ants of the chip business.

There are more than six billion jellybeans in operation in the world today. They are everywhere in our daily lives – automatic doors, cars, household utilities, remote controls, cameras, TVs, smart cards, satellites circling the globe recording all sorts of things. Jellybeans cal-culate when you should take your car for a service. They work out when to harvest wheat fields, or when to order more spare parts for factories. They manage the brakes in your car, measure temperature and humidity in bathrooms, switch electricity on and off at any given time in machines; they register whether a door is open or closed, whether a currency is running or not, and whether a light bulb is broken or not. They sit in Formula One racing cars, where they send continuous information to the DAGs (data acquisition geeks) at the pit stop, who then use computers to collect information on how the car is doing. They sit in soft drink vending machines to keep track of the number of cola cans left, and pass these data to the can distributor and even to the thirsty, Internet-connected student down the hall. They are, in other words, anywhere and everywhere, each of them pulsing away relentlessly, always ready to perform its tiny, designated role in our lives. And we will have more of them in the future – in our shoes, clothes, driving licences, personal health cards, car tyres and footballs, where they will measure and respond to everything from heat to moisture, wind, technical status, temperature and movement.

While larger chips might be likened to frozen thoughts, the jellybeans are frozen reflexes. They check things and communicate responses or send instructions for action, which are followed by machines or noted by humans. They transmit trillions and trillions of instructions every second of the day, some of them for people to react to, but the large majority from one man-made device to another. As they become even smaller and cheaper, they can fit into even more places – under our skin, and on any consumer goods we buy. There are 10 trillion objects manufactured each year, and there will be quite a few places to place them.

The most fascinating thing about jellybeans is the dynamics that evolve when you have many of them around. Let's just go back to biology for a minute. Ants appear to be rather stupid. However, it is impressive what they can achieve when they all work together. The same applies to jellybeans. Jellybeans often work in swarms, and in doing so they seem to achieve things that appear far more amazing than what each of them could do individually. The magic effect comes when we enable many of them to stay aware of each other, and when we enable information to flow laterally rather than just from the centre outwards, or from the periphery to the centre. The network beats the hierarchy.

Just imagine for a moment a supermarket where there are jellybeans attached to each object for sale. The jellybeans could manage the prices – not price lists, but the price label on each and every item. Assume, for instance, that the supermarket has a real-time, online tracking of inflow and outflow of goods for sale. It would consequently also have electronic records of the numbers of goods sitting on stock (in the warehouse and on the shelves). The central computer could now use artificial intelligence to predict how many units of each item would be sold at each hour of the day. It would learn, for instance, how much of a certain product people were likely to buy during each hour of the day, each day of the week. It would also learn to adjust for holidays, long weekends, etc. Now add some jellybeans that measure the outside temperature, and the computer would now be able to learn how demand for beverages and charcoal fluctuated with the weather conditions. Then add jellybeans that read electronic television programme guides, and the computer would know that the demand for beer would increase before a major sport match.

So what would the jellybean attached to an item do? Well, firstly it would adjust the price continuously to optimize the revenues. Perhaps it would be aware of the age of the produce and implement a drop in price as it started getting a bit old. Who knows, perhaps someone would invent a jellybean that could sense the maturity of a cheese. Or estimate it by sensing temperature and calculating time since manufacture? There is no end to the creativity.

What we are building with chips – large and small – is something that adds up to much more than just an artificial ant system. There are thousands of different jellybeans and thousands of software programs and chips for smarter computing. And billions of electronic connections tying them together. There exists, for instance, a system that enables airplanes to exchange observations with each other, and to use that information for optimizing travel paths. Car manufacturers have developed systems to track the status of the cars electronically in order to detect common errors and even alert a driver before an error happens. And the healthcare industry is evaluating how it can insert chips that monitor the health of people with life-threatening diseases and alert them and their doctors when trouble is ahead.

Even refrigerator manufacturers are working along the same lines. Put a computer inside the fridge equipped with an automatic barcode reader that will automatically tell us when the milk is getting old, and what we have in the fridge to be cooked for tonight. In marketing terms, you could advertise a product at the central spot of consumption, for example a cola ad right in your face when you open the fridge door (let's hope it can be switched off . . .).

Communication between the chips reduces traffic jams, saves energy, and secures a safer and more pleasant journey when we travel. Although we do not have a master plan for it, and one could never be made, we are in reality now building an artificial ecosystem, which is growing in size and diversity at an astonishing rate. There will be millions of man-made silicon 'species' – millions of different chips and computers working in conjunction with each other. And as the digital mass explodes, so will the digital diversity. The system is self-organizing, because this is the only way that mankind can build something that is complex and evolves very quickly.

Digital Emotion

The Internet changes everything. Most business executives will agree with this. There are very few industries that are not directly or indirectly affected by the Internet, which is why some equity analysts say things like 'There will be no more talk about Internet companies in a few years from now, because all companies will be Internet companies by then. Those who didn't make the change will be out of business.' Internet banks, Internet retail, Internet purchasing, Internet recruitment, Internet cell phones, Internet cars (this seems particularly interesting: perhaps they will give away the cars and make money on the services?), Internet travel booking, Internet home working, Internet whatever. How about Internet broadcasting?

Well, how about it? Despite the fact that broadcasting is all electronic, it isn't a part of the Internet. Not if you receive it on your TV or radio anyway. Not even if you receive it on a desktop, since the delivery over the Internet is based on the Internet's TCP/IP protocol, which is not a broadcasting technology. (Every stream over the current infrastructure of the Internet is transmitted individually to each receiver. Broadcast, on the other hand, streams out once to all users). But broadcasting will change. Here are some of the things we can expect:

- Digital broadcast channels (the next generation of digital television) will contain hot spots that you can click on to receive background information, shop or whatever.
- Your PC, your mobile phone and your car entertainment system will, when turned on, receive constant streams of high-resolution video, music, software downloads, etc.
- Leading Websites will contain TV-quality video and DVD-quality music whenever it matches the editorial theme. A news site might, when breaking news is taking place, switch to live, high-quality video coverage on a part of the screen.
- You will have music sites that sound as good as a normal radio and have no download time, which will supply you with related background information.
- Financial information will be delivered in guaranteed real time, and will combine live stock quotes with video interviews from analysts, live reports from the trading floors, etc.
- Games sites will have games that are linked to real, live events. A flight simulator might, for instance, be connected to live weather data, so that you can simulate a flight through a real typhoon as it evolves; or you could race against real cars in a computer animated car race.
- You will be able to follow sports events ('goal scored!') as a video clip on your mobile phone delivered automatically within seconds after the event took place.
- You will have real-time holographic video, where you will be able to sense the three dimensions of the filmed scenario.
- Digital entertainment will be beamed directly into your retina from your glasses so that you will not even need a screen to see it with.

All this will be possible for two reasons. The first reason is that broadcasted content will be encoded in IP (Internet protocol) format and will be transmitted through dedicated broadcast 'speed lanes' over land lines and through airwaves to all devices with Internet addresses. The broadcasted content will then appear as a part of the Web browser environment, integrated seamlessly and without the performance degradation and download times that we normally associate with the Internet. As this power becomes reality, video will become an absolutely natural part of our life no matter where we are. We will have streaming video to our wristwatches, mobile phones, glasses, cars, fridges, etc.

The second reason that these new services will be possible is that bandwidth is exploding. George Gilder, the famous technology futurist, wrote an article in *Forbes ASAP* in December 1994, which contained a surprising prediction: Fibre optics, he said, had the potential to carry 25 000 000 000 000 bits per second down a single strand. This was 10 000 times more than what most engineers expected to be the ultimate barrier of optical bandwidth. Gilder based his prediction on the assumption that two emerging technologies would be implemented:

- the erbium-doped amplifier, which keeps optical signals pure and strong over long distances;
- wavelength division multiplexing (WDM), whereby two or more optical signals of different wavelengths could be transmitted simultaneously in the same direction over one fibre, and then separated by wavelength at the receiving end.

Gilder wrote an interesting follow-up article ('Fiber Keeps it Promise') in *Forbes* (April 1997), where he elaborated:

> . . . the intrinsic capacity of every fiber line is not 2.5 gigahertz. Nor is it even 25 gigahertz, which is roughly the capacity of all the frequencies commonly used in the air, from AM radio to ka-band satellite. The intrinsic capacity of every fiber thread, as thin as a human hair, is at least one thousand times the capacity of what we call the 'air.' One thread could carry all the calls in America on the peak moment of Mother's Day. One fiber thread could carry 25 times more bits than last year's average traffic load of all the world's communications networks put together: an estimated terabit (trillion bits) a second.

What is the end result likely to be? The Internet will reach us in many different places, and will combine the speed and emotion of television with the smart bits of a computerized network. It will always be on, will contain full-motion video and DVD-quality sound, will provide software access on subscription basis, and will be much faster than what we know today. By combining two-way communication with broadcasting to create a rich and seamless experience, the Internet will bring emotional content to people in an environment where they can re-spond directly. This emotional content can be anything from a market-ing person motivating his or her salespeople to the broadband entertainment for millions. And emotion is, as we know, an important part of marketing.

Freedom

The first generation of computing was what we now call Master–slave computing. The focal point was the central computer handling instruc-tions, a concept that was reflected in the central thinking organizational structure. Then came the slightly more pleasant sounding Client–server computing, where smaller computers were connected to the centre to handle information centrally as well as locally. This coincided with decentralization of organizations. This was followed by networked computing, which stimulated collaborative work practices and enabled computers to handle relationships beyond the limits of the internal corporate network. The ultimate demonstration of this phenomenon happened when the scientists from the SETO programme, which col-lected signals from outer space to look for signs of radio transmission from intelligent life forms, made a smart decision. They needed more data processing power than they could afford, so they asked volunteers to use their PCs to work on their tasks while they were otherwise running idle. The scientists managed, within a short period of time, to sign up two million private users, who all allocated their PC processing power to the task.

The next big wave is *ubiquitous computing*, which sets us free to seek information and execute digital tasks wherever and whenever we prefer. The concept of ubiquitous computing was first described by Mark Weiser, who worked for the Computer Science Laboratory at Xerox PARC. In 1988, Weiser started talking about a new paradigm in computing, where computers would be everywhere and where they would become more and more integrated into the landscape until we would reach the point where the world would be the interface. He said:

Inspired by the social scientists, philosophers and anthropologists at PARC, we have been trying to take a radical look at what computing and networking ought to be like. We believe that people live through their practices and tacit knowledge so that the most powerful things are those that are effectively invisible in use. This is a challenge that affects all of computer science. Our preliminary approach: activate the world. Provide hundreds of wireless computing devices per person per office, of all scales (from 1" displays to wall sized). This has required new work in operating systems, user interfaces, networks, wireless, displays, and many other areas. We call our work 'ubiquitous computing'. This is different from PDAs, dynabooks, or information at your fingertips. It is invisible, everywhere computing that does not live on a personal device of any sort, but is in the woodwork everywhere.

Weiser's approach was new. We have all seen old science fiction movies portraying a future where we would be surrounded by incredibly intrusive technology – giant computers filling rooms. Weiser's vision was the opposite. He imagined that computers would become *less* dramatic:

For 30 years most interface design, and most computer design, has been headed down the path of the 'dramatic' machine. Its highest ideal is to make a computer so exciting, so wonderful, so interesting, that we never want to be without it. A less-traveled path I call the 'invisible'; its highest ideal is to make a computer so imbedded, so fitting, so natural, that we use it without even thinking about it. (I have also called this notion 'ubiquitous computing', and have placed its origins in post-modernism.) I believe that in the next 20 years the second path will come to dominate.

Things are going in Weiser's direction. Computer designers working with ubiquity are now using the term 'calm technology' to describe computers that get less and less dramatic until they, like electric motors, will become invisible to your eyes and ears. Figure 2.4 shows the four big waves of computing.

Invisibility is not the only aspect of the ubiquity idea. It is also about availability. The Internet is now available via a multitude of different network infrastructures. Content can flow over telephone lines that have been made much faster through a technology called digital subscriber line (DSL). Or it can flow via cable TV networks, TV satellites, wireless TV networks (called 'digital–terrestrial'), mobile phone networks, public wide area networks (WANs), corporate local area networks (LANs), or even one day over our own body area networks (BANs).

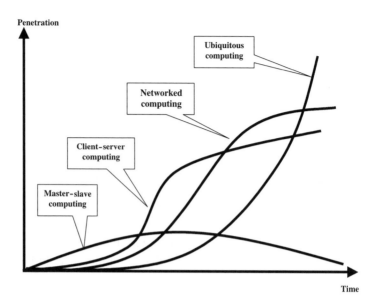

Figure 2.4 The four big waves in computing. Computing has evolved in four big waves, from master–slave computing to client–server, networked and ubiquitous computing.

Body area networks? Our bodies are actually great electric conductors with lots of salt and water. A BAN would keep personal information very personal except, when you decide otherwise. It could transfer information from your tennis shoes to your watch to tell you how much energy you were burning, or your ID information to the handle of your office doors, or your electronic business card to anyone when you shake their hands.

Ubiquity is provided not only because data can flow over many different networks. It is also because it can flow to many different devices. To PCs, of course, but also to television sets, mobile phones, personal digital organizers, car entertainment systems, and even to refrigerators with plasma screens. And to numerous devices equipped with jellybeans. There will soon be hundreds of categories of devices that connect electronically.

One of the key solutions will be a technology called 'Bluetooth'. Bluetooth was a Danish Viking king who convinced many wild tribes to work together (a rare skill, which is also in demand in electronic communication markets!). It is also the name of a short-distance wireless communication technology (a set of technologies, in fact), which is poised to become one of the leading solutions for home entertainment networks as well as for much of the interdevice communication in offices and elsewhere. The initiative was originated by Ericsson, Intel, IBM, Nokia and Toshiba, but more than 2000 other manufacturers had, by mid-2000, joined the family. Arguably, this makes it the fastest growing industry standard ever.

Any device that uses the Bluetooth standard for communication will contain a tiny microchip incorporating a radio transceiver. This will enable the device to communicate without the use of cables. The communication will facilitate fast and secure voice and data transmissions, even when the devices are not within line of sight. The radio operates in a globally available frequency band, ensuring compatibility worldwide. A device equipped with Bluetooth will be able to do all manner of things, such as:

- *Simplified access to external networks.* This is done by recognizing and connecting to different types of networks through a Bluetooth connection. For instance, you can just as easily and instantly connect to the Internet via a mobile phone as via any Bluetooth-enabled wire-bound connection.
- *Cable replacement.* The technology eliminates the need for troublesome cable attachments. You can, for example, send and receive emails on your mobile computer via your mobile phone, even when the devices are not within line of sight.
- *Personal ad hoc networks.* All your Bluetooth-enabled devices can be set up so that they automatically exchange information and synchronize with one another. For instance, if you accept an appointment on your handheld device, the appointment is automatically accepted in your desktop PC as soon as the two devices are within range of each other.

The Bluetooth technology is designed to be fully functional, even in very noisy radio environments, and its voice transmissions are audible under severe conditions. The technology provides a very high transmission rate and all data are protected by advanced error-correction methods, as well as encryption and authentication routines for the user's privacy.

The following (provided by the consortium) are examples of what the technology will enable:

- *Mobile phones.* Your gateway to the world. In the Bluetooth world, your mobile phone is automatically connected to other digital devices. You could, for instance, connect your mobile computer to the Internet and send and receive emails, even when the linked devices aren't within line of sight. The mobile phone becomes your gateway to the world.

- *Same phonebook everywhere.* For practical reasons, we tend to have one phonebook in our mobile phone, one in our mobile computer and/or desktop computer at our office, and one in our handheld device. With the Bluetooth technology, all of these phonebooks can be instantly and automatically synchronized each time the devices come into range. This makes it possible, for instance, to key all phonebook updates into your computer and then easily and swiftly transfer them to your mobile phone – or vice versa.

- *Same phone for all needs.* By making your office intercom system Bluetooth-enabled, and adding a Bluetooth-compatible base station at home, you can use the same phone wherever you are. At the office, your phone functions as an intercom with no telephony charge. At home, it functions as a portable phone with a fixed line charge. And when you're on the move, the same phone functions as a mobile phone with a cellular charge. The switches will be made automatically, depending on which network within reach is available.

- *Mobile computers.* Most of the information you need on a day-to-day basis is stored in your mobile computer. Therefore, it is also your natural information warehouse. With the Bluetooth technology, all connections between your mobile computer and other Bluetooth-enabled devices are instant and automatic. You can send files from one mobile computer to another as easily as over a LAN. Or you can surf the Internet regardless of your location – through a mobile phone or any Bluetooth-enabled wire-bound connection. Sending and receiving email and faxes is just as simple. As long as your mobile computer is within reach of a Bluetooth access point, you have a fast, secure and wireless connection to the outside world.

- *Handheld devices.* All the information stored in your handheld device will be accessible from other Bluetooth-enabled devices – without connecting them by cable. And vice versa. Important records, such as your calendar, contact list, phonebook and to-do list, will never have to be outdated in any of your devices, regardless of where and when the information was entered. With your handheld device connected wirelessly to a mobile phone, or any wire-bound Bluetooth-enabled connection, you can actually send and receive email, notes and simple documents by tapping on the screen. The connection is established automatically and will be maintained even if the devices aren't within line of sight.

- *Headsets.* Keep your hands free for more important tasks. The wireless headset is as easily connected to a mobile phone as to any other Bluetooth-enabled wire-bound connection. Through the headset, you can automatically answer incoming calls, initiate a voice-activated dial-up and end a call, providing your mobile phone is equipped with proper functionality. You can also transfer a call or alter the ringing tone between the headset and your mobile phone or a mobile computer. The wireless headset offers very high-quality sound and allows audio playback from a mobile computer. You can also control volume and microphone with the headset from a computer or mobile phone.

- *Office equipment.* The Bluetooth technology connects all office peripherals wirelessly. You can, for instance, connect your desktop or mobile computer to printers, scanners and faxes without ugly and troublesome cables. And you can increase your sense of freedom in your everyday work with a wireless connection for your mouse and keyboard to your computer.

- *Cameras, still image and video.* The possibility to transfer still images and video clips between camera and a mobile computer is a good example of the versatility of the Bluetooth technology.

When your digital camera is Bluetooth-enabled, you will be able to send instant postcards as still images or video clips from any location by wirelessly connecting your camera to your mobile phone or any wire-bound connection.

- *Other electronic devices.* The potential of the Bluetooth technology is virtually unlimited. One after another, new applications and products, as well as increased functionality, will be introduced. Small handheld scanners, portable hard disks, wristwatch information centres, refrigerators, coffee machines and presentation projectors are just a few examples of where fast and secure wireless connection will simplify our everyday life. As for all other Bluetooth applications, all connections are instant and automatic and maintained even if the devices aren't within line of sight.

- *Internet access and email.* The Bluetooth technology brings the world to you regardless of your location. From your desktop computer, mobile computer and handheld device you will always have fast and secure wireless access to the Internet – through a mobile phone (cellular) or through a wire-bound connection (PTSN, ISDN, LAN, xDSL). To connect is as simple as switching on the lights.

- *Office LAN.* By installing a Bluetooth network at your office, you will eliminate the troublesome and ugly cable attachments and redefine the meaning of connective flexibility. No longer will you be bound to certain locations for connections or have to install new cables for new workstations. Each Bluetooth-enabled unit can be connected to more than 200 other devices. And since the technology supports both point-to-point and point-to-multipoint connections, the maximum number of simultaneously linked devices is virtually unlimited. Furthermore, whether you're working in mobile mode or back at the office, your mobile computer can be connecting automatically to the LAN.

Bluetooth is just one of many examples of technologies that enable networking capabilities in new products. The consequence of all these devices and telecommunication between them will be that we can do online tasks from virtually anywhere. This, again, will further stimulate the growth of virtual worlds. A virtual world is an electronic replication of something that also exists in the physical world:

- *Virtual reality.* Goggles and sensors in your clothes enable you to feel like you are moving around in a remote location.
- *Virtual store.* Simply e-commerce.
- *Virtual ballot box.* You vote from home, or your mobile phone, or whatever.
- *Virtual exhibition.* You can see the exhibition booths and presentations from a desktop.
- *Virtual job.* You work for a company, but never show up. Just use the web.
- *Virtual alien.* You can live in Bangladesh, but to all practical effects work for an American company without having a green card.
- *Virtual village.* An electronic community where people from many different places are able to find each other easily.
- *Virtual sports game.* You can participate in somebody else's real game without actually showing up.

The global village is really coming. So is global vision.

Global Vision

One of the unique effects of digital networking is the ability to see what couldn't be seen before. You can see places and people far away and see things that haven't, in reality, been created yet.

Just think about manufacturing. The old-fashioned way of building a physical construction was to make drawings, then build a small or simplified physical model, and then go for the real thing. But more and more physical objects are now designed with a computer, so that you can experience it, and even test it virtually, before you manufacture it. The manufacturing process can be linked directly to the virtual design. This process is called CAD/CAM (computer aided design/computer aided manufacturing). A new term is now being coined: 3D print. Why? Because most of the time, when we transfer something from a screen to a physical output, we do it by hitting the 'Print' button. The design goes from virtual world to physical paper. A 3D print is really just taking that thought a step further. A 3D print could be a wind surfer, or a car or even an aeroplane (the 747 was the first Boeing to be designed entirely on computer).

Digital networks are also providing vision across distances. Furniture manufacturers, for instance, are installing cameras so we can view their furniture on their website. This can work in many ways, for example a man walks into a furniture shop and finds a sofa that he likes. He calls his wife from his mobile and asks her to view the website so she can see it as well. Accessing the webcam, she directs it via the Web to point to the sofa, where her husband is sitting. She likes it as well, and the purchase is made.

Webcams can be put anywhere: shops, roads, tourist resorts, airports, social gathering places, even your own house. A webcam provides us with knowledge to help us decide how to spend our time and money.

The car industry is probably one of the best opportunities. Just think about choosing from an online catalogue your favorite car, seat, leather colours – within seconds. The automated IP server calculates the price, goes to the manufacturing system, allocates your car slot, and informs you when you can watch online your car being manufactured. Remember the storytelling role we spoke about in Chapter 1?) It's no longer just the product and the service you buy. It is the story and the ability to get involved.

Digital cameras are often fixed to special locations. They can, for instance, sit on satellites orbiting the earth while constantly recording and transmitting global images. But they can also be mobile, for example a digital camera that you hold in your hand, or perhaps, one day, a camera so small that it is built into our glasses or clothes, or a broadband mobile phone that we could use to see each other no matter where we were. So, with all these digital eyes everywhere, why not ask someone to lend you an eye? A new company, Remote-I, offers exactly this. Remote-I has developed a network whereby you can ask anybody anywhere on the globe who has a digital recording device – camcorder, third-generation mobile phone, whatever – to record whatever you need to see and beam it to you over the mobile phone network or the Internet.

The result of all this is, perhaps more than anything, increased freedom. We don't have to be at a desk to connect. The Internet will be in the air, linking into the things we wear and use everyday. It will enable us to connect to any man-made intelligence, benefit from swarm effects, and share emotions wherever we are.

Individuality

The simplest way for a company to find out what the market wants is to use market aggregation strategies. Market aggregation is reminiscent of the military principle, reconnaissance by fire. This is a very simple means to finding out where the enemy is. Simply fire in all directions and expect the enemy, if he is out there, to fire back and thus disclose his position. A company that aims in all directions will not only hope to find enemies, but will find out where there is positive interest in its products and services.

A more sophisticated marketing approach is market segmentation. This concept was introduced to the marketing literature through an article published in 1956 by Wendell R. Smith, 'Product differentiation and market segmentation as alternative marketing strategies'. This method has also been used by the military. In a classical soldier-to-soldier battle, the power of each army is, in principle, related to its weapon strength multiplied by the number of soldiers. However, in modern battles armies try to focus their efforts so that they first overwhelm one part of the opponents forces, then move on to the next part. Lord Nelson used this principle at the Battle of Trafalgar, where he first surrounded and defeated one half of the Spanish-French armada before attacking and defeating the other. In the 1970s, Canon used this strategy when they entered the British copy machine market that Rank Xerox dominated. Canon focused initially only on Scotland. Not until it had reached a market share of 40% in Scotland did it approach the London area.

However, market segmentation is not the ultimate approach. Developed in 1956 as a part of the managerial school of thought, it didn't assume the presence of widespread digital networking. The digital economy has brought a new approach called 'mass customization' or 'one-to-one marketing' (some spell it 'one2one'). The idea is to use digital technology and electronic networking to enable every single customer to tailor design their own versions of the product. There are plenty of examples:

- Order a car with your preferred configuration over the Internet, e.g. Toyota.
- Some hotels register your personal preferences, so next time you check in, your room is tailored to your needs, e.g. The Ritz Carlton chain.
- Some e-commerce sites greet you with special offers that reflect past purchase patterns and the taste space you appear to belong to, e.g. Amazon.com.
- Order a PC with your preferred configuration over the Internet, e.g. Dell.

It is probably fair to say that Michael Dell and his people were the leading commercial pioneers of this concept, as they were the first to build a global business empire based entirely on mass customization. By doing so, they enabled customers to configure their PCs exactly the way they wanted them, and they were still able to deploy industrial mass production methods to manufacture these machines.

One of the results of mass customization is being able to overcome the 'hard middle' problem. The 'hard middle' is a phenomenon in retailing referring to the groups of consumers that are too small to be viewed as a worthy target segment, but too expensive to serve individually. Mass customization makes it possible to serve these smaller user customer clusters.

Another result is that the producer, through the digital ordering process, can obtain extremely valuable information about what customers really want – and when they want it, who they are, and where they live and work.

The third result is that the customer becomes involved directly in the production process, not only through the digital product ordering process, but also through their digital communication on related bulletin board discussion groups. More and more frequently, the manufacturers of a product will discover that they have customers out there who know more about their product and its use than the people who actually manufacture it, and that some of these customers care deeply about the quality and have great ideas for improvements. It is no wonder that more and more software companies release free beta versions of new software so that they can generate a flood of comments from potential users of the final release. Communication with the 'hobby tribes' can become an integrated part of the development process.

The fourth result of one-to-one marketing is that expensive misconceptions are prevented. We can call these 'demand illusion'. Imagine that you run a recording company and you are considering which jazz records you wish to print as CDs. Your editors choose one and it is distributed to retails shops, where people are looking for old jazz records to buy. The information coming back to you is that there is a healthy demand for this particular artist, and so you choose to produce more of the same. However, the intermediary (the retailing shop) does not produce an efficient flow of information back to the manufacturer. You get false information showing apparent demand, but what you don't see is that there would be a much larger demand for a CD edition by another artist – a comparable situation to where a manufacturer produces a number of units, say cars, in a specific colour, and the car sells only because that is what the retail has to offer.

The last, and perhaps most important, effect of one-to-one marketing is enhanced freedom. You get what you want, when you want it. You are not forced to choose between product configurations that a marketing expert tried to guess that you would prefer. You get what you *really* want or need, and you can order it on Sunday evening, just before midnight, if that is a good time for you. Furthermore, you can learn about yourself if the network you connect to is smart enough to recommend things that other people in the similar taste space have opted for.

Market Spaces

When things are in radical change, the initial public reaction is often radical as well. People often fall into two camps – those who are wildly optimistic and those that are aggressively against. The 'naives' and the 'don't get its'. For example, when the Internet took off around 1994, opponents wrote the whole thing off as a play toy for geeks and naive teenagers. 'How can you make money from something you give away?' The supporters, on the other hand, saw computers taking over numerous business tasks, including a complete elimination of the distribution chain and many other sectors. They predicted that anybody in retail should consider some pretty urgent career planning; because there would no longer be anything between the producer and the final user (other than package delivery services).

Neither group was right. You *can* make (big) money on something you give away. But there *will* still be many intermediates between the producers and the users of products in the future. Perhaps even more than there has been in the past. But the mediators will work in completely new ways. One of the first to describe this phenomenon was the futurist Paul Saffo, who sits on various advisory boards, including AT&T Technology Advisory Board, the World Economic Forum Global Issues Group, and the Stanford Law School Advisory Council on Science, Technology and Society. Saffo wrote in a 1999 article:

> Barely five years ago, the notion of commerce over the Internet was anathema. Later, Internet commerce became the hottest thing in cyberspace. Once Net-commerce became real, conventional wisdom held that the Internet would spell the death of advertising.
>
> In fact, the Internet has turned out to be a huge new advertising frontier.
>
> Now a much more dangerous bit of conventional wisdom is on the loose. It is the notion that information technologies will bring about disintermediation – that is, networks and information systems will allow buyer and seller to interact directly, thereby eliminating intermediaries and radically shortening value chains.
>
> There's only one problem with this theory. It's directly at odds with what is actually happening. Rather than eliminate intermediaries, information systems do quite the opposite. Information systems are powerful commercial tools because they lower transaction costs. Lower transaction costs enable new kinds of transactions, which lead to new market niches and, overall, make the market environment more complex. In short, information systems create openings for new intermediaries to discover and occupy.
>
> Meanwhile, old intermediaries are disappearing, but that is only part of the picture. What seems to be disintermediation is really a mirage, one static piece of a larger process in which the introduction of new information systems disturbs market

environments, creating slots for new intermediaries whose presence threatens older established intermediaries who either disappear or adapt to new market realities. What looks like disintermediation is but one frame of a larger dynamic of 'Disinter-remediation'.

Other than inventing one of the most complicated words in marketing ('disinter-remediation'), Saffo managed with this article to encapsulate a key development in the markets: while the new mediators don't have to operate retail shops, there are a number of tasks that they actually do fulfil. So, we could perhaps even try to extract yet another rule ('Saffo's Law', of course) from his writings: The diversity of distribution services is reversibly proportional to the cost of distribution transactions (Figure 2.5).

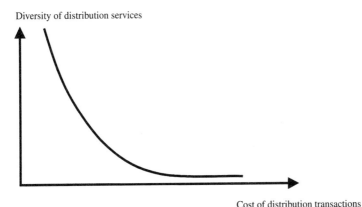

Figure 2.5 Saffo's Law. Saffo's law describes the relationship between the average cost of distribution transactions and the diversity of applications provided by distribution intermediaries.

Digital communication technology reduces the cost of distribution, therefore we see that the distribution intermediaries in digital economies (the sectors that many thought would disappear) are alive and multiplying in numbers and diversity, as they provide the following:

- product comparisons and ratings
- adjust prices to market conditions
- search and filtering facilities

- new product alerts
- access to product-related discussion forums and chat sessions
- recommendations based on users' search and purchase habits
- electronic payment services
- shipping services
- electronic product presentations
- product-related information, such as news, regulatory data, research, reviews, awards, etc.
- information about the buyers' previous purchases, wish lists, shopping baskets, etc.
- digital exchanges linking any buyer to any seller within product categories
- real-time data for advertisers.

They will also, in most cases, have to maintain a physical inventory of goods, and they will need real people for customer support. Lots of them, in fact. Experiences from electronic stock market trading services indicate, for instance, that you need a customer support person for every 50 active customers.

So, the real change is not disintermediation but a vast improvement of the services that intermediators provide to their customers. More and more goods will be traded in ways that are reminiscent of electronic stock markets. Customers will have full transparency, prices will change more frequently, and transactions will be possible from anywhere.

Timelessness

People have, over the last centuries, increasingly been forced to divide their time according to strict rules. As individuals, we have time at the workplace, time to go shopping, time to relax, and time to have fun. Corporations have time to do annual plans, time to do stock listings, time to submit orders for supplies. But our relationship with time is changing in three ways:

- we are moving from a crude time-slot paradigm towards more flexibility;

- we are moving from a batch paradigm to a real-time paradigm;
- we are moving towards an always-on economy.

The issue of flexibility is particularly important for the individual. We are used to a paradigm where we have to be at the workplace at given times, shop at given times, go on holiday at given times. While some of these limitations will remain, the trend is towards a much more flexible life. More and more people will be able to work and shop from anywhere, and do it at whatever is the most convenient time for them.

The real-time aspect is mainly relevant to the way corporations run. Nature generally works in real time, responding to input the moment it is sensed. It was the industrialized society that made it necessary to introduce discrete planning and action intervals. However, with electronic networking, we are able to move back to the more efficient and natural approach – the real-time corporation.

The always-on model relates to a change from switching-based telecommunication towards packaged-based communication. We shall elaborate on this a bit later, but let's just explain briefly now: switching means that you reserve a communication line when you are online. Package-based technology means that you are online all the time, but you share the bandwidth with others. You use not the whole line, but only what is actually necessary to transmit the digital packages.

Funny Money

Who in this world is allowed to create money? The obvious answer is central banks. Anyone else who makes money should go to jail. Right? Wrong. Economists define money as 'any item simultaneously serving as a unit of account, a medium of exchange, a store of value, and a standard of deferred payment.'

Within this definition, it became clear to economists at the beginning of last century that commercial banks actually also create money, since credit is, to all intents and purposes, the same as money. But banks are not the only ones doing this. Any company can issue money. Frederich Hayek, the Austrian economist, wrote in his *Denationalization of Money* of 1976:

> Money does not have to be legal tender created by governments. Like law, language and morals, it can emerge spontaneously. Such private money has often been preferred to government money, but government has usually soon suppressed it.

Here are some examples of Hayek's 'private money':

- *Electronic telephone cards.* You pay an amount for a smart card, which you can use in a public payphone. The card works until you have used up the prepaid amount. The prepaid idea can also be used on standard telephone bills.
- *Cyber loyalty schemes.* Websites give you cyber points for the time you spend there. These can be used for buying different things.
- *Casino money.* You buy chips, which you use when you gamble.
- *Travellers' cheques.* You buy cheques, which are guaranteed by the issuing institution.
- *Events smart cards.* Organizers issued, for instance, 300 000 smart cards at the 1994 Atlanta Olympics.
- *Prepaid smart cards.* Used for TV set-top boxes, where viewers can use content, pay per view or games.
- *Frequent flyer miles.* These are equal to money restricted for use on specific items.

Let's look a bit closer at the frequent flier miles as an example of money issuance. Can this really be compared with money when you can only use the 'miles' to buy flight tickets? Perhaps not. But what about when the 'miles' can be used for renting cars, booking hotels, flying with a large number of alternative airlines, even 'buying' a new camera? Clearly this is a situation in which an alliance of corporations jointly issue and redeem money. So, they need a monetary policy. Expansionary policy means issuing a lot of money. This will boost current sales. However, the more money a company issues, the more it will cannibalize future sales.

The concept of corporate money is still in its infancy, but it will grow and grow, mainly through the phenomenon of e-money. E-money can be divided into two categories:

- *Identified e-money* contains information revealing the identity of the person who originally withdrew the money from the bank. Sophisticated versions may contain the person's DNA signature or a digital picture.

- *Anonymous e-money* works like paper money: once it is with-drawn from an account, it can be spent or given away without leaving a transaction trail.

There is also another important distinction:

- *Online e-money* can be spent only by interacting electronically with a bank.
- *Offline e-money* can be spent without having to directly involve a bank.

And there is a third distinction:

- *One-off e-money* is the 'electronic purse'. It's basically a travellers' cheque that gives exact change. You pay a specific amount for a card that goes into vending machines. When the card is debited down to zero, you throw it away.
- *Rechargeable e-money* is the smart card that you can re-charge as required.

E-money has many advantages:

- *Transnational.* A company can make e-money legal tender, irre-spective of the country it or its customers are in.
- *Smart way to build loyalty.* E-money can be used to generate a network loyalty amongst the companies participating in a corp-orate monetary system.
- *Creates user convenience.* E-money is more secure than cash (through, for instance, fingerprint identification and identity codes), it fits better into vending machines, you don't have to worry about change, and you can obtain and use it anywhere.
- *Facilitates payment processes in retail shops.* Infrared payments will, for instance, not require any physical movement of notes and coins.
- *Prevents vandalism.* There will be no notes or coins in vending machines.

There are also disadvantages, though. Corporate money makes it more difficult for central banks to control money supply and for tax authorities to collect taxes. If a person earns 100 000 dollars, 40 000 frequent flyer miles, 6000 website bonus points and 4000 Microsoft Bills, What should he or she pay in tax?

OPEN STANDARDS AND CONVERGENCE

Standards is the phenomenon in which an industry or society agrees on some common rules for the way that certain things will and do function, for example cars drive on the same side of the road and respect red lights. Standards can be frustrating at times, but lack of them tends to make things worse (try driving in Cairo).

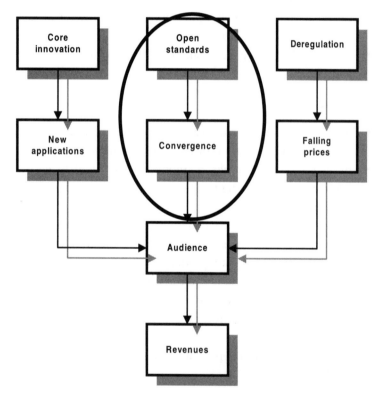

Figure 2.6 Open standards and convergence in the digital economy. Open standards and convergence are keys to the development of large, digital communities.

Standards in the digital economy is largely a software issue, since software makes the whole thing work (at least some of the time). And yet, the software business wasn't really a sector in its own right in the early days of computing. In the early days, hardware vendors made software – and it worked only with their own hardware. Computing consisted therefore of a number of isolated, incompatible technology islands, and once you had chosen a vendor, you were pretty well stuck with that company.

However, around the early 1950s, the first serious independent software companies emerged, and the industry has kept growing ever since, developing into three main categories: mass-market software, enterprise solutions and professional software services. By the late 1980s, it became clear that the greatest independent software companies could become extremely profitable and valuable, which led to the boom in standards. Since then, the use of standards like WAP, GPRS, DVB and UMTS in the telecommunication industry has exploded. In 1999 alone, 780 standards were filed and agreed upon within the European Telecommunication Standards Institute (ETSI), and this is expected to exceed 1000 very shortly.

Software and other technology standards come in two forms: proprietary and open. A proprietary standard is usually owned by a corporation. Companies that control proprietary standards will typically not let external companies get full access to the code, and they can change it at will. Somebody who wants to inspect some software, for instance, needs access to its source code (a description of what its different parts do). It is incomprehensible without this code.

An open standard is a published definition that is possessed by no one and may be used by anyone. It is typically managed by either a consortium of companies or by an institution. DVB – the standard for digital video broadcasting – is, for instance managed by the International Broadcasting Union. HTML (the Internet standard) is managed by the World Wide Web Consortium. The important issue about open standards is that anyone has access to use it, provide input to it, criticize it, and get full insight into how it works and how it is expected to evolve. A special case of the open standard is the 'social group standard'. This is a standard driven by individuals rather than companies or organizations. A great example is the operating system Linux: people with little or no commercial interest have collaborated over the Net and, in the end, built a solid operating system. Another example is Open Source, which brings applications at little or no cost to the computer world. These two new paradigms in marketing software packages and related services will have a major influence on the future success of very large players, such as Microsoft and Oracle.

The widespread use of open standards has three key effects in the market:

- *Companies move from vertical integration to virtual and horizon integration.* Without open standards, a company in the TMT (technology, media and telecommunications) sector would typically try to control all related technologies for its customers (vertical integration). With open standards, they are more likely to focus on doing a few things extremely well on behalf of many different markets (horizontal integration) and then tie that in with complementary products and services through alliances (virtual integration).
- *Increased division of labour.* Any delivery to any final user involves input from a higher number of suppliers in an open standards economy.
- *Increased creative pressure.* Open standards enable substantially more players to make applications and to have a meaningful chance of getting them to the market.

The proliferation of open standards in digital technology has been instrumental in its aggressive growth. The Internet, for instance, is built on a series of open standards, as are digital broadcasting and telephony. Telephony is a good example, because while Europe and most of Asia/Pacific chose early on to endorse a single standard for mobile phones (GSM), the approach in North America was to let several competing standards develop. This became a key reason why mobile phones were deployed much faster in Asia/Pacific and Europe than in the USA.

AOL and the Inefficiencies of CompuServe and Prodigy

A good example of how fast fortunes can change in the digital market place is the example of AOL, Prodigy and CompuServe in 1995. People had more or less declared Steve Case, the founder of AOL, as a losing party and the company's future did not look promising. CompuServe and Prodigy were the winning brands and most financial analysts praised their successful business models, which centred around their own 'walled gardens' where they defined the services they wanted their customers to see and spend time with. However, the consumers clearly knew what they wanted – to go outside the walled gardens of CompuServe and Prodigy. AOL saw this coming, and quickly adapted by letting other content suppliers participate and become part of the AOL brand. Few could have imagined in 1995 that AOL would acquire Time Warner, the prestigious and well respected global media conglomerate, four years later. AOL succeeded in understanding this new get-me-all content economy. As we are writing this book, AOL has over 24 million customers. This is driven by the network effect, where AOL is protecting its own instant messenger service (IMS) to its own subscribers but is offering a free download to other ISP customers. These non-AOL customers download the AOL IM program free of charge, which gives AOL access to its competitors' subscribers. This is a smart way of marketing: offer the service free of charge, therefore making no profit, but by the network effect, which is so important in the digital economy, the competitors' customers slowly but surely migrate to AOL.

Core Convergence and Derived Application Diversity

It would be natural to assume that the increased use of open standards came at the price of reduced creativity. After all, standards and creativity sound like opposing worlds. But such an assumption misses an important point: standards lead to convergence. Convergence means that an application can be used in many different contexts. The Internet, for instance, can be described simply as a convergence phenomenon. Online computing and digital telecommunication existed in numerous forms before the Internet, but these were largely not interconnected. Once it became clear that the Internet would really be the core standard for telecommunication, everybody started to adapt. This convergence made it feasible for thousands of new companies to develop an avalanche of great applications. So, standards and convergence stimulate an increase in the number of applications. The more standardization, the more applications (Figure 2.7).

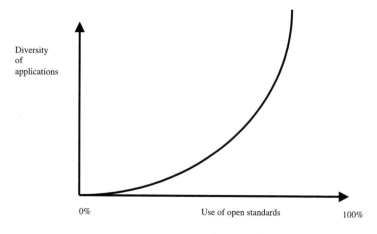

Figure 2.7 Open standards and diversity of applications. Use of standards doesn't mean fewer applications. It means more. Open standards tend to become adapted widely and thus lead to an explosion in the number and diversity of applications.

The Internet was just an example. The convergence continues. Voice, television and two-way data exchange are all converging around open standards. Cars in the future may contain communication systems that combine global positioning systems (satellite systems that tell the car where it is on the map) with digital terrestrial broadcast (a television broadcast infrastructure that can also be used for the broadcast of interactive data) and UMTS (third-generation telephony) to create the ultimate, seamless mobile media solution.

DEREGULATION AND FALLING PRICES

One of the most important drivers of the technology boom has been a wave of global deregulation. This process, which began in the USA and was later adapted by the European Union (EU), has now swept across the globe. It is:

- breaking down custom barriers;
- opening up competition in industries such as banking, stock and commodity exchanges, airlines, public transportation, utilities and telecommunications;
- reducing the administrative burden on numerous industries;
- allowing foreign ownership and cross-border mergers of companies in almost any industry.

The combined effects of these developments have been increased competition and innovation on all fronts. Any company that stands still in a deregulated world, even for a short period of time, will immediately face aggressive competition from numerous other players – start-ups as well as giant global players. Management in formerly protected industries have thus faced a new situation in which they have had to act aggressively just to maintain their market positions. One of the most interesting outcomes of this has been a consistent pressure on prices for anything related to digital networking (Figure 2.8). Prices of hardware, connections and bandwidth have all come tumbling down.

USER COMMUNITIES AND REVENUES

We have now looked in some detail at the avalanche of new applications for digital networking, the continued convergence of core technologies, and the consistent fall in unit prices of hardware, software,

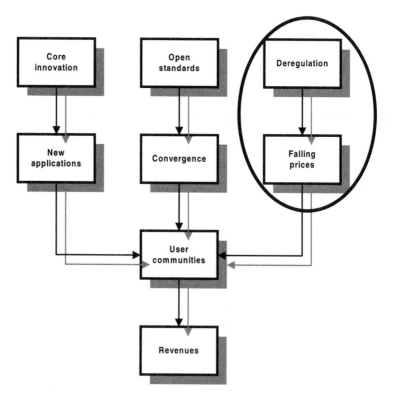

Figure 2.8 Deregulation and falling prices in the digital economy. One of the key drivers of the digital economy is the consistent downward pressure on unit prices. Deregulation contributes considerably to this pressure.

digital content and bandwidth. The combined effect of these trends is an explosion in the size of the user communities, which increases overall revenues, even as unit prices fall (Figure 2.9).

However, this is not the end of the story. What we have described so far might look like a system with predominantly linear dynamics. The input of the model is core innovation, open standards and de-regulation, and the result is increased audience and revenues. There is a linear relationship between input and output. Double the input, we might think, will mean doubling the output.

But this is wrong. The reason that companies operating in digital economies can develop in such spectacular ways is that digital

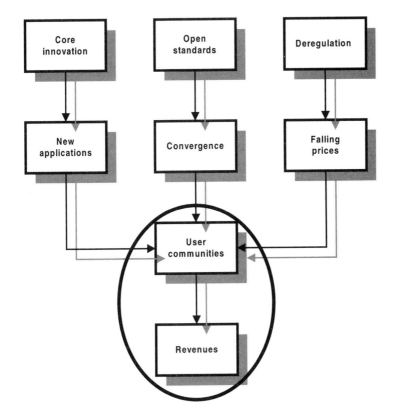

Figure 2.9 User communities and revenues in the digital economy. The digital revolution has created a growth in uptake of new products and services that has never been seen before.

economies contain very strong nonlinear growth properties. These nonlinear properties are based on feedback processes, or feedback loops.

FEEDBACK LOOPS

A feedback loop is a process in which the result of a chain of events has an impact on earlier stages in the same chain. There are two main classes of feedback loops – positive and negative. Positive means that things escalate automatically after ignition; negative means that they are dampened. Assume, for instance, that your house catches fire. The fire generates heat, evaporation of gases, and air turbulence, which together stimulate more fire. This is a positive feedback process – an escalation process. The fire also triggers a call to the fire brigade, who show up and extinguish the fire. This is a negative feedback loop.

Economic systems, old and new, are full of feedback loops. Most explanations of the business cycles and irrational financial markets are, for instance, rooted in understanding the numerous positive and negative feedback loops.

The most important feedback processes in the digital economy are illustrated in Figure 2.10 by the highlighted arrows pointing upwards. Let's briefly explain how each of these feedback processes works:

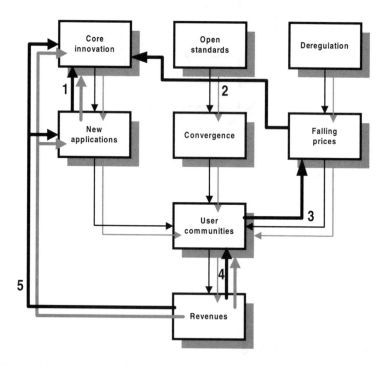

Figure 2.10 Positive feedback loops in the digital economy. It is essential to understand the role of feedback loops in order to understand the new economy. It is the loops that create the most dramatic economic opportunities and threats for all companies.

1. *New applications stimulate core innovation.* We need technology to create technology. A company that produces innovative solutions for digital networking will depend on huge amounts of technology to get on with its tasks. The tools needed could be anything from new measuring instruments to software object that make it possible to write highly sophisticated programs without having to deal with every single line of code.

2. *Falling prices stimulate core innovation.* If you work in high tech, you will have noticed that everything you need, from phone lines to PCs to high-end computers, just gets cheaper and cheaper, which increases your ability to innovate within a given budget.

3. *Growing user communities lead to falling prices.* The more units a company sells, the lower the average unit price. This is true for any company that has high fixed costs and low variable costs, but particularly for companies operating in the digital economy. Some even find that their variable costs are virtually zero, enabling them to reduce prices dramatically (or even to zero).

4. *Growing revenues stimulate growth in user communities.* This is a question of wealth. Growing revenues create added wealth in society, which enables more people to connect to digital networks.

5. *Increased revenue leads to new applications and core innovation.* The bigger the audience connected to a digital network, the bigger the commercial incentive to create new applications for them.

So what we have is a system where a series of positive feedback loops lead to continued, inherent escalation. This observation is vital to the understanding of four phenomena, which perhaps best describe the essence of what the new economy means for the marketing strategist:

- network effects
- minimal marginal costs
- path dependency
- leverage.

We will now examine these escalation phenomena more closely.

Network Effects

Eric Schmidt, Chairman and CEO of Novell, once made an interesting statement about products and communities: 'I don't need a finished product. What I need is a special movement. I need to build a community of players who will help develop the offer, who will refine the language, who will join together to make this happen.'

Novell makes products and yet the CEO didn't need a product. He needed a movement. Why? One of the key reasons is the phenomenon of 'network externality', which can be defined as 'a change in the benefit, or surplus, that an agent derives from goods when the number of other agents consuming the same kind of goods changes'.

The term network externality sounds very academic, so practical people prefer to talk instead about 'network effects' (also because 'externality' ceases to be the correct academic word if a company manages to 'internalize' the effect, meaning taking active measures to benefit from the externality). 'Agent' sounds a bit strange too, but it's how economists refer to people or companies.

Network effects can be explained through an example. Imagine you are the only person in the world that owns a fax machine. It's useless to you. Five years later, there are millions of other fax machines around, so you can start sending and receiving faxes. The value to you of your fax machine has risen because other people also bought similar machines. The same would apply for phones, Internet connections, or anything else that communicates.

The fact that something becomes more valuable if many people have it is relatively new in economics. It certainly doesn't apply to horses, money, antique paintings, gold, land, mines, hotels or fancy cars.

The leading thinker on network effects has been Robert Metcalf. Metcalf learned in 1970 about a network called AlohaNet, which was used for data communication between the Hawaiian Islands. The network was very simple. Content was divided into small packets, each of which had a digital header containing information about where it should go to, and a payload, which was the actual content. The key feature of AlohaNet was that you did not have to wait for a dial tone when you wanted to send something to someone. You could send your message at any time, and each of the digital packages would then go on its way towards the final user; once it got there, the final user's computer would send back acknowledgement information. If your computer never received the acknowledgement information, then it could conclude that the digital package was 'lost in the ether', as Metcalf expressed it, and your computer would then just send the package again. This concept is today called package switching, as opposed to the dial-tone concept, which is called circuit switching.

The AlohaNet was based on a smart idea, but the implementation was not perfect. Metcalf decided to change that and developed a series of mathematical algorithms that would enhance the performance of such a network considerably. His discovery was called Ethernet, and it came to dominate corporate networking around the globe, as he founded 3Com Corporation. However, while working with commercial networking implementation, it dawned on Metcalf that the benefit of any user to be connected to his networks was really in exponential proportion to the number of other connected users. This observation has since been termed 'Metcalf's Law', and it is one of the most important keys to understanding the economics of digital societies. Some of the most important consequences of network effects are:

- *They invite companies that may benefit from them to invest heavily in development of their network.* A potential network effect may be so important to a company that it chooses to invest more money in supporting the external environment than in its internal affairs.
- *They can lead to the success of an inferior technology.* An inferior technology could be the first to be launched commercially and generate a network effect. It may then be very difficult for a subsequent supplier of perhaps a much better technology to enter the market.
- *They make the battle for standards central to many businesses.* Businesses can succeed or fail simply by being on either the winning or the losing side of a battle for standards.

Network effects are very attractive to tap into. Once you have them, your business tends to feed on itself, escalating almost on its own. But the escalation patterns are not simple. Let's take a look at a few examples. The first one is development of a digital network infrastructure, which is mainly driven by the supplier's resources, for instance, the limited availability of people to carry out the physical installation (Figure 2.11).

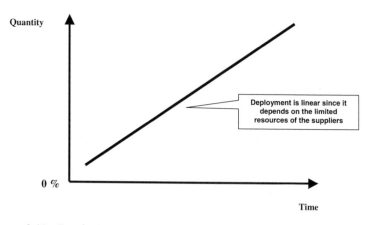

Figure 2.11   Developing a network effect that depends on limited supply resources.

The second case involves deployment of networking technology in a world where the relevant supply resources are cheap and plentiful. The penetration pattern here is more likely to take the shape of the graph in Figure 2.12.

Launching a technology that is supposed to enable network effects can take huge amounts of time. The Scottish inventor Alexander Bain invented the basic technology for the fax machine in 1843, and AT&T introduced a wire photo service in 1925, but faxes remained niche products until the early 1980s. Likewise, the first implementations of the Internet existed in 1970, but it didn't take off commercially until 1994. By 1998, you seemed doomed if your business didn't have an Internet strategy, and now you might as well be dead if that strategy is not implemented.

Networking technologies appear dormant for such a long time before taking off because a technology that becomes more valuable to

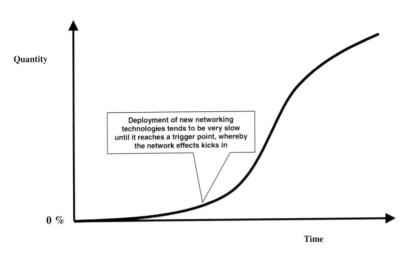

Figure 2.12 Developing a network effect where supply resources are cheap and plentiful.

any single user as other users connect is likely to have little or no value when there are few users connected. So, in the early days there is a huge barrier to growth. What happens is often a sort of 'ketchup effect' – the technology is dormant for many years until it suddenly explodes, just like when you hit the ketchup bottle 10 times without getting any ketchup out, and then suddenly the whole bottle empties on your plate.

It can also be difficult for companies offering new commercial services over an existing electronic network to establish a network effect. The most attractive of these businesses tend to be exchanges – stock exchanges, or exchanges of any kinds of goods and services. But how do you get people to join an exchange where there is virtually no trading going on? Markets with network effects are winner-takes-all markets, and one of the key challenges for any company operating in these markets is to figure out how much they have to invest in order to get to the trigger point that starts the positive feedback loops. Invest too little, and nothing happens; invest too much, and the company might go bust before it happens.

It is also a challenge to see how the network effect will work once it has started. The mobile phone industry has provided good examples.

Many analysts assumed in the early days of this industry that we would see aggressive growth penetration rates until reaching about 30% penetration. The growth rates would then level off. The opposite happened. It seemed that about 30% penetration triggered some new dynamics so that the subsequent penetration growth actually *accelerated*. Forecasts were upgraded from saturation around 30–50% to saturation at more than 100% – meaning that the average citizen would eventually have more than one mobile phone. This reflected various studies (see, for instance 'At home in the universe' by Stuart Kauffman) that showed that as the number of connections among elements in a system rises above half the number of elements, the probability of cascading events rises dramatically. This represents a phase change in the system, resulting in strong, nonlinear responses to the external inputs (Figure 2.13). In other words, when the network penetration reaches a certain trigger point, it goes bananas.

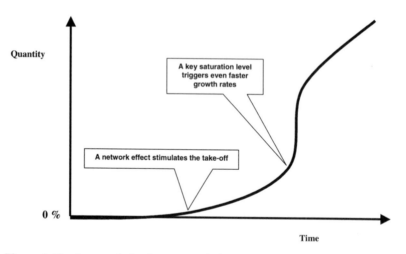

Figure 2.13 A network development with dual trigger points where the dynamics change.

Minimal Marginal Costs

The second key characteristic in the overall behaviour of the new economy is the increasingly common experience of minimal marginal costs. A business has fixed and marginal costs. A car manufacturer incurs the fixed costs of developing the cars and the brand, and building the factories. The marginal costs are the costs of producing and selling each car.

Companies in the digital economy use the same distinction, but often the marginal costs are very small. Often, it costs virtually nothing for the supplier to replicate the product once it is developed. Just think again of the software business. It costs very little to copy software. Even if you have to invest a billion dollars in developing a new software package, it costs you only a few cents per unit to distribute it over the Internet.

The rational response is to drive prices of the digital products down in order to create large communities, network effects and the business escalation that it brings.

Path Dependency

Path dependency is a well-known phenomenon in many sciences. It is, for instance, a key element of the chaos theory, when it is called 'sensitivity to initial conditions'. Biologists tend to call it 'contingency' as they use it to describe the irreversible character of natural selection. Path dependency is the determination, and perhaps lock-in, by small, insignificant events. Path dependence in economics assumes that minor initial events in the development of a market can lead to massive chain effects and feedback loops that have very significant effects at a later stage.

The leading writer on path dependency in the digital economy is Brian Arthur, a scientist with experience as Dean and Virginia Morrison Professor of Economics and Population Studies at Stanford University, Citibank Professor at the Santa Fe Institute, and a Coopers & Lybrand Fellow. Arthur and others have divided the phenomenon of path dependency into three different categories, where the perspective is the degree of damage to free competition that path dependency may represent:

- *First-degree path dependence* is one that does no harm to the economy in general. This means that initial, perhaps insignificant, events put us on a path that cannot be left without some cost, but that this path actually happens to be the best one.

- *Second-degree path dependence* is not so harmless. It assumes that information is imperfect when the first event occurs, and that this leads to decisions that are not efficient in retrospect. We have no way of knowing that we are moving in the wrong direction from the beginning, but sensitive dependence on initial conditions leads to outcomes that are regrettable and costly to change. However, we cannot say these events are inefficient because we had no way of knowing this at the time.
- *Third-degree path dependence* is really bad. This is the situation where the market goes in the wrong direction, even though information about the error is available when it happens.

There are numerous stories about cases of path dependency in the new economy, where companies have used superior marketing techniques to gain lasting leads over companies with superior technology. As software people say, it's the marketing gods that make the software kings. Software kings build on path dependency.

Leverage

Companies operating in the digital economy can use three basic forms of leverage:

- Financial leverage to fund the initial stages of growth.
- Commercial leverage to extend the reach.
- Branding leverage to extend the demand.

Let's look at financial leverage first. Digital economies tend to have increasing returns (which is great), but they also have huge start-up costs, so a company needs to fund what are initially loss-making activities. New-economy companies tend to use two main forms of financial leverage: issuing *stock options* and issuing *shares*. Both have the advantage that although the company has to pay back a huge return if things go well, it is not obliged to pay anything back if it doesn't work out. Stock options are promises to give the employee large earnings in the future if things go well. This option enables the company to retain key talent without paying exorbitant salaries (this is not the only reason that stock options are used, though). Shares serve the same purpose, but by providing cash rather than saving costs.

In order to obtain financial leverage, a company has to put a value on its activities. In traditional, well-established businesses, this is typically done by benchmarking key financial indicators of the company against the sector norm. Investments look at price/earnings (the total valuation of the company's shares divided by its earnings), or combined annual growth rate (CAGR), or book value (depreciated value of corporate assets). The situation is different in the digital economy, however:

- Demand for very aggressive growth tends to lead to larger initial losses.
- Network effects and increasing returns create large uncertainty about the future growth and earnings (difficult to estimate CAGR and price/earnings).
- Book value is a mostly irrelevant benchmark, since new-economy companies seek to produce maximum returns with minimum use of assets. Furthermore, new-economy companies will often write off most assets immediately after they are purchased.

This has led to new approaches. One of the leading valuation guidelines in new economy start-ups is the discounted cash flows (DCF) model. This is based on discounting expected future cash flow with a risk factor and adding a residual value at the end of the simulation period. While DCF is the dominant, common valuation basis, the actual funding process is different at various stages of the corporate growth cycle:

- *Seed phase.* In the very early stages, influential 'angels' provide prominence and networking (relationships) to the company. Seed-phase financing is often based more on evaluation of the business concept and the management than on any written budget forecasts.
- *Venture phase.* This is the point at which the company has a well-developed business plan and future scenarios can be imagined

with a minimum of clarity. The company now needs to attract partners that bring additional relationships and business deals that stimulate the network effect.

- *IPO (initial public offering).* This is when the company goes public and raises additional capital in the process. It can exert influence over who is allocated shares, but it will not control whether those people sell the shares shortly after or not.
- *Post-IPO.* After the IPO has been completed, the company loses control over who its investors are. However, this is offset by the fact that its stocks are turned into a currency, which can be used for the acquisition of other players in the market, so that it can be consolidated for maximum network effect.

Financial leverage is important, but it is far from sufficient as a tool to develop critical mass and network effects. The company needs *commercial leverage* as well. This is obtained through cooperation with networks of people and companies that interact within the company's business sphere. The issue here is that a company trying to build a network effect needs to mobilize forces that are far beyond what it can do through its own resources. It must thus build a series of partnerships involving everything from joint product developments to participation in standards setting bodies, joint distribution, comarketing and integrated sales efforts. (See Appendix A).

Branding leverage is equally crucial, but sometimes overlooked. It is great to produce a product and stick your brand on it, but it is far better to be able to stick your brand on somebody else's product. When you enter a Boeing plane, for example, the first brand you notice might not in fact be that of Boeing but rather the prominent General Electric branding on the engine.

THE ESSENCE OF FEEDBACK IN THE NEW ECONOMY

One of the consequences of the strong feedback processes in the new economy is that companies have to be driven more aggressively. We have to gamble, because trying to be very careful is even more dangerous than trying to be very aggressive. It's like being hunted by animals in a heavy fog: dangerous to run too fast, but even more dangerous to run too slowly. Use leverage and invest aggressively in the development of your network before someone else beats you too it. Time flies, and managing against time has become one of the key skills of winners in the marketing game. The next section of the book is thus devoted to this aspect.

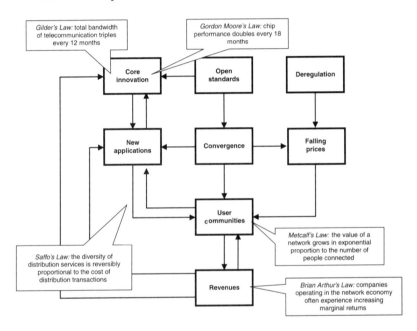

Figure 2.14 Essence of feedback processes. The digital economy is dominated by very powerful, positive feedback processes. The diagram shows how some of these serve to accelerate different processes in the overall dynamic system.

TIME

'Speed is the form of ecstasy the technical revolution
has bestowed on man.'

Milan Kundera

CHAPTER 3

TIME

One of the world's most impressive companies, General Electric, embarked in the early 1960s on a project to investigate why some of its units were much more profitable than others. The project showed that the most successful units were typically those with a high market share. The project was later turned over to the Strategic Planning Institute, a nonprofit research organization. Here, the information was developed into the Profit Impact of Market Strategies (PIMS) database, and the institute soon began covering thousands of strategic business units within hundreds of companies. The results continued to suggest that high market share is one of the keys to high profitability. The reasons were clear:

- Buyers will often feel that it is safer to buy from a leader.
- During downturns, distributors often reduce their inventories by eliminating the smaller suppliers.
- Leaders can afford to hire the best (and most expensive) management.
- Leaders have economics of scale in their production and marketing efforts.

The new economy has created additional reasons:

- Leaders will enjoy the biggest network effects.
- Leaders will benefit most from the minimal marginal costs
- Leaders will enjoy path dependency.
- Leaders will be best positioned to utilize leverage strategies.

It is often easier to capture a large market share during the early growth phases of the market, when distribution patterns are still fluid and crucial potential partnerships have not yet been made.

THE ECONOMY OF SPEED

So how do we achieve the highest market share? Literature based on the traditional marketing schools has listed numerous factors, but one of the most important has been to *start before the competition*. Jack Trout and Al Ries noted in their article, 'Positioning cuts through chaos in the marketplace' (1972) that: 'You can see that establishing a leadership position depends not only on luck and timing, but also upon a willingness to "pour it on" when others stand back and wait.'

They mentioned typical examples:

> When you trace the history of how leadership positions were established, from Hershey in chocolate to Hertz in rent-a-cars, the common thread is not marketing skill or even product innovation. The common thread is seizing the initiative before the competitor has a chance to get established. In someone's old-time military terms, the marketing leader 'got there firstest with the mostest.' The leader usually poured in the marketing money while the situation was still fluid.

So, traditional marketing schools observed the importance of moving first because it could take a company to the point of being the biggest. However, the digital economy adds a twist to this. It is now not only important – it is often absolutely crucial. New-economy companies feel that they have to move fast – and they act accordingly:

- It took Microsoft 10 years to reach 100 million dollars in revenues;
- AOL took nine years;

- Yahoo! took five years;
- Onsale took four;
- Amazon three;
- Priceline spent just one and a half years reaching 100 million dollars in revenue.

Every aspect of the way companies are managed is accelerating. We use email, pagers, mobile phones and websites for instant, ubiquitous communication. Companies announce their products before development has even begun, then prelaunch them when they are half finished. People raise money to start companies for markets that don't exist yet, and they go public before they have had their first profitable quarter.

We also change the whole work culture to optimize for speed. More and more decisions are delegated, and more and more formal processes are changed from batch mentality to continuity. The winning company is the one that is closest to real-time management. The traditional management culture just isn't competitive anymore. Either you lead the road, or you become a part of it.

INCREASING RETURNS

The reason for all this is that the digital economy is full of companies that have the potential to enjoy the pleasant phenomenon of increasing returns. Increasing returns are the consequences of a combination of network effects and minimal marginal costs. To understand this, we can start by looking at the traditional supply and demand curves that we were taught in business schools in the old days before the Internet (Figure 3.1).

In Figure 3.1, marginal returns are *diminishing*, typically because traditional companies depend on limited resources of input factors. The formal expression for this is 'diminishing returns'. Think, for instance, of an oil company. An oil company will first drill for oil where it doesn't cost very much, perhaps in an Arabian desert. It makes a bundle on this business. However, if it wants to find more oil, it has to look at

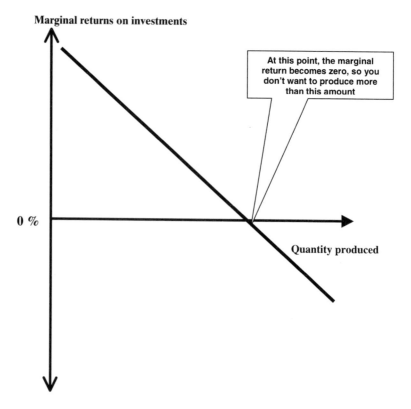

Figure 3.1 Diminishing returns in traditional industries. The classical assumption in the old economy is that the company is utilizing limited resources so that the marginal returns on investment decline as volume is increased.

more remote places, perhaps, say, deep down at the bottom of the sea in the Antarctic. And if it wants to find still more, then it has to deploy very complex technologies to squeeze the last drops out of each field. The returns on each new drilling project are lower than the previous. The same goes with farming – first the farmer cultivates the best land, and then he or she moves on to less lucrative fields.

A market with *increasing returns* works completely differently (Figure 3.2).

This situation occurs if a company has very large start-up costs but very low marginal costs. The effect becomes even stronger if there is a network effect.

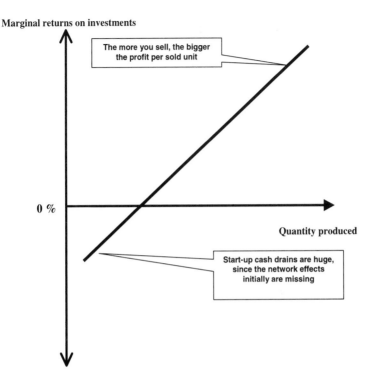

Figure 3.2 Increasing returns in the digital economy. The typical situation in the digital economy is that marginal returns get larger and larger as the company grows.

Brian Arthur has studied numerous examples of positive feedback and increasing returns in the economy, and he has looked at how increasing returns could lead to the magnification of small, random events (path dependency). His main interests are the economics of high technology, the new economy and how business evolves in an era of high technology, cognition in the economy, and financial markets. Arthur has defined increasing returns as 'the tendency for that which is ahead to get further ahead; for that which loses advantage to lose further advantages.'

If increasing returns is important in business, then speed of execution becomes vital in marketing. It becomes a key part of the marketing mix.

SPEED IN MARKETING

Any businessperson knows roughly what the term 'marketing mix' means. It is the combination of choice of products, price, promotion and place that a company deploys. Jerome McCarthy, in his book *Basic Marketing: A managerial approach* (1960) called it 'the 4 Ps of marketing'. Software people have developed an alternative model:

- PPP = PowerPoint presentation
- PP = Preproduct
- P = Product

What they refer to is the way software companies launch products. First, they talk about it madly, having only PowerPoint presentations to show. Then they kind of launch it, even if it is only half ready. This is a soft launch of a preproduct. Finally, you get the real thing.

The reason for this approach is that time is everything in the digital business. We need to inform customers and scare competitors by pre-announcing far in advance that we are coming up with something. (This strategy is often ascribed to IBM, which used it as it came under pressure in the mainframe market.) TMT (technology, media, telecommunications) companies are completely focused on the time dimension, so in the new economy, we need to add the fifth P, which stands for 'pace':

- product
- price
- promotion
- place
- pace.

Pace refers to the crucial time dimension. We live in the economy of speed. We can order custom configured computers and receive them hours later, or custom configured cars, which arrive by the end of the same week (in Japan, at least). Product development cycles are collapsing, and everything is moving closer and closer to real time as we install chips and networks everywhere. Brands such as Pokemon and Palm Pilot grow from nothing to global madness within a few years.

One of the key concepts that makes this increased pace a possibility is electronic data interchange (EDI). EDI links computer systems between suppliers and their customers for purchase orders, invoices, billing, and record keeping. Many people think of it as a way of saving resources – and it is. But the biggest effect for many companies is that it enables real-time response to the market conditions. It is really the backbone system that connects swarms of computers and chips with each other and with the people that depend on the output.

This is where exchanges of services and products, both in the supply chain and in the demand market, will radically change. One just has to look at the new exchanges/purchasing communities founded recently for the aircraft, chemical, pharmaceutical, car manufacturing industries etc. These exchanges are all built on a real-time economy in which the company that can present itself best in the digital world will run with the rewards.

Just as in elite sport, the winner is a huge star, number two is OK, number three is so-so, and number four – what was the name . . .? In business, the first company in any sector to enjoy the benefits of significant increasing returns can crush its competition.

Second-mover advantage model

First-mover advantage is often everything in the technology/ digital-driven economy. But only if the first mover gets it right.

In the late 1940s, people started looking at jet-powered passenger planes. The first one was the Comet from De Havilland in the UK. This incredible plane could fly at 450 miles per hour, well over twice the time speed of classical propeller planes. Comet began commercial flights in May, 1952 and everybody at that time thought De Havilland, with its first-mover advantage, had an incredible future. Unfortunately, De Havilland did not get a successful ride – a year after the first commercial flight, De Havilland had to report a string of mid-air crashes that killed all passengers on board. One plane got lost over Italy, one over Elba, and one in India. Boeing then introduced their first jetplane, the

707. So, in this case, the innovator – De Havilland – did the research, and the follower – Boeing – made it big.

Another great example is from the mobile phone industry, where Dancall in Denmark pioneered products based on GSM. Despite its brilliant concept, the company did not understand how to ride its own market innovation, whereas Ericsson and Nokia did.

The picture is mixed, as the table below illustrates:

Product	Innovator	Follower	Winner
Diet cola	R.C. Cola	Coca-Cola	Follower
Pocket calculator	Bowmar	Texas Instruments	Follower
Instant camera	Polaroid	Kodak	Innovator
Microwave oven	Raytheon	Samsung	Follower
X-ray scanner	EMI	General Electric	Follower
Fibre-optic cable	Corning	Many companies	Innovator

VALUE

'There is more to life than increasing its speed.'

Mahatma Ghandi

CHAPTER 4

VALUE

W e have looked at the technologies and the time dimension that drive the digital economy. The next natural step is to investigate where and how the market players can extract commercial value out of these environments. Where is the money? Is there any?

WHAT IS STRATEGIC VALUE?

There are many complex definitions of commercial value in the literature, but we will stick with a simple one: commercial (or financial) value is the *return on investments*. It is dividends or any capital appreciation of a company, whether it is through share price increases on public stock exchanges or other liquid markets, through acquisitions, or by other means. Such commercial value will directly benefit anyone with a financial interest in the company. It will benefit any investor or option owner, such as employees, public investors, management, board, venture capital funds, pension funds, or whoever. This commercial value can be created in two ways:

- brilliant execution
- strategic positioning.

You can make shareholder value through brilliant execution alone, even if your business has a dull or outright difficult market position. Just assume that employees in your company work twice as hard as those in a competing company. This will generate commercial value. However, while brilliant execution is a fine and rare art, it is largely tactical or operational. It is a question of motivation, organization, communication and numerous other disciplines that we think are outside the scope of this book. Strategic positioning, on the other hand, is a question of how you position a company to succeed in the digital economy.

HOW DOES STRATEGIC VALUE OCCUR?

We all know about Adam Smith (1723–1790), the Scottish economist who wrote *The Wealth of Nations*. Adam Smith believed that an 'invisible hand', would automatically restore balance in an economy that had been through a shock. The invisible hand was the result not of any central policy or management effort, but of all the decisions made by millions of individuals as they sought to maximize their own welfare.

Smith's ideas have had a bumpy road (the biggest bump provided by Karl Marx), but they experienced an academic revival in the 1940s and 1950s with a new 'neoclassical' school (Karl Marx had called Smith and his immediate successors 'classical economists'). This school developed largely because of mathematical discoveries: different economists had learned to create models that could simulate how a large economic system behaved, simply by applying mathematical rules to individuals' behaviour. In order for the models to work, each individual was assumed to be basically rational. They were 'intelligent agents' who never got mad or did anything stupid, and were not wasteful. And because the agents were intelligent, the markets would always 'clear' – prices would always be set at the level where buyers and sellers could agree, whether it was for labour, capital, goods or services. It was often assumed, for instance, that:

- there was perfect competition;
- everyone had perfect knowledge about products and prices;
- there were no geographical barriers;
- there was perfect labour mobility;
- there were no changes in technology.

The beauty of these models was that we could change the assumption on the micro level, i.e. the rules of how each individual could behave, and the models would then calculate what this would mean to the entire system. But this was not realistic: the economy doesn't consist entirely of intelligent agents and markets do not *always* clear. Just think of the Great Depression in the 1930s: there were lots of people willing to work and lots of products looking for buyers. But the clearing mechanisms failed, because real markets had queues, bottlenecks and huge inefficiencies. They had pricing that led to a temporary build-up of excessive inventories or unemployment, and they had products that didn't succeed commercially, simply because people weren't aware of them.

But what would happen if markets were perfect? How could anyone make a profit in a market with absolutely perfect competition? They couldn't. Ironically, the system would, if we imagined that it could be brought to work seamlessly, actually beat itself. Perfect competition would remove any incentive to invest, and the economy would grind to a halt.

The point is that strategic commercial value is generated by a company that is able to pursue a market inefficiency. The bigger the inefficiency, the bigger the potential gain. This pursuit may, in some cases, contribute to a closing of the opportunity gap, but the combination of increasing returns and path dependency in digital economies means that the pursuit of opportunities of inefficiency can be rather long-lasting and highly lucrative. We can call them 'value zones'.

CLASSIFYING VALUE ZONES

Market inefficiencies have always existed. There have always been trade barriers, lack of visibility, geographical distances, monopolies, cultures and numerous other constructions that leave pockets of inefficiency where companies could have a ball. The biggest advantage of the digital economy is that it improves the market clearing process by providing visibility and removing friction. The most important effect of widespread use of very smart computers is that it brings us closer to the ideal scenario that economic models sometimes assume. The computers will be the efficient agents that can cover up for their less rational creators, to create a system that is more efficient.

But, as many inefficiencies are eliminated, others have evolved. There are, as we mentioned earlier, four key sources of market inefficiencies that dominate the digital economy:

- *Network effects*. The ability to increase the value of a product or service as more and more people use it.
- *Minimal marginal costs*. The tendency for marginal costs to approach zero in digital economies.
- *Path dependency*. The tendency for first movers to establish leads that are virtually impossible for late starters to catch up on.
- *Leverage*. The ability to mobilize company-external resources that far exceed the company's own resources.

Any value strategy designed for the digital economy will benefit from at least one of these phenomena. However, since many of the most attractive value models in the digital economy benefit from several or all of these phenomena, we will divide the models according to another formula. We shall classify them into five major categories (Table 4.1).

Asset-based Value Models

VBThe principle of an asset-based value is to invest in the creation of an asset that generates a sustainable competitive advantage. There are five leading asset-based models:

- brand-building model
- disruptive invention model

Table 4.1 Major value model

Major value model categories	Specific value models
Asset-based value models	Brand-building model
	Disruptive invention model
	Build-to-be-bought model
	Blockbuster model
	Standards model
Network-based value models	Community-building model
	Platform and standards model
	Digital exchange model
	Profit multiplier model
Timing-based value models	First-mover advantage model
	Organized-for-speed model
Customer relationship-based value models	Secondary sales model
	Mass customization model
	Central aggregator model
Cost-based value models	Critical scaling model
	Low-cost business design model
	Consolidation play model

- build-to-be-bought model
- blockbuster model
- profit multiplier model.

Brand-building model

If people like and trust a brand, then they are more likely to buy the product, and they are also willing to pay more for it, sometimes a lot more. So it's no wonder that creation of brands has always been one of the central themes of marketing. The authors of the commodity school would, for instance, typically describe the brand as an integral part of the product. Much of their research was related to the shopping experience – when, how, and why a person would decide to make a purchase.

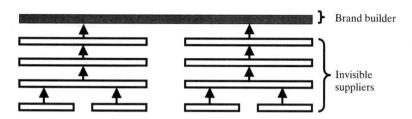

Figure 4.1 The brand-building model.

The perception and awareness of the brand was often seen as the key differentiation. Branding was also central to the buyer behaviour school, where in particular Kuehn (1962) inspired an avalanche of research into brand-buying behaviour.

The branding exercise has become even more important in the new economy, where more and more of what we buy is immaterial, complex and continuous. How do we decide which software to buy, or which mobile phone service agreement to have? We cannot open the product and inspect the interior, nor can we look into the future to see how the product will be serviced. So we place our faith in the most trusted brand.

Branding is a managed phenomenon. It is the skilled supplier that builds great brands. This is often done through excellent brand leveraging, where a company manages to brand a phenomenon that extends far beyond its own deliveries. When we buy a mobile phone from a service operator, for instance, it is typically the brand of the handset manufacturer that we see and associate mostly with, rather than the brand of the service provider.

Branding leverage can be achieved through cobranding, joint branding, or simply by placing your brand on a screen image, a sound stream or a piece of hardware that you didn't make yourself. When you buy a PC, you will probably see the Intel brand on it. Intel made the chip, but it is also cobranding the whole box.

The ultimate success in branding is achieved when a company wins the association with the whole product category, for example people have, for many years, called photocopiers by the brand name –

'Xerox'. Today, when we say office automation, we think Microsoft; when we say servers, we think Sun Microsystems; databases, Oracle, and so on.

Disruptive innovation model

A traditional route to creating asset values is to make an innovation that totally changes the game in an industry. This can be a technical innovation (anything from the PC to the carving ski) or it can be a completely new way to do business – a new business process. Several authors have written about how such innovations reach markets, and how good they need to be in order to succeed on their own merits. The role of innovations has, for instance, often been described by authors from the buyer behaviour school, who looked at how innovations typically penetrate the market. A classic book regarding this subject is *Diffusion of Innovations* by Everett M. Rogers (1962).

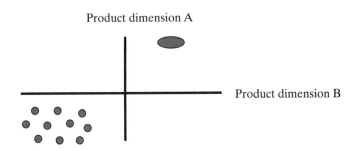

Figure 4.2   The disruptive innovation model.

Peter Drucker made an important contribution to this thought process when, in his book *Innovation and Entrepreneurship* (1985), he described what has since been coined 'Drucker's Law'. The law stated that a new innovation had to be at least 10 times better than what it replaced in order to overcome the sheer marketing muscle and production capacity of existing producers.

Build-to-be-bought model

It is often said that the Internet industry is scary because eventually there will be only a small fraction of the thousands of Internet companies left. The car industry is frequently used as an example: in the early 1920s there were thousands of car manufacturers, but by the end of the twentieth century, only a few were left. We assume that the rest went belly up, but this is not the whole truth. Many of the smaller manufacturers were bought up, and their shareholders were able to walk away with substantial returns on their investments.

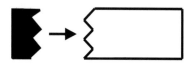

Figure 4.3   The build-to-be-bought model.

Numerous companies that are founded in the new economy are created by engineers and operated with a total focus on technology, rather than marketing. Such companies cannot succeed on their own, but they can be very attractive acquisition targets for larger companies that have the marketing machine, but perhaps not this particular technology or expertise. Some of the start-ups that are bought in this way may be run by management that are unaware of the importance of marketing, but others may be very conscious of the strategy – they are essentially built to be bought. A stellar example is Broadcast.com, which was acquired by Yahoo for about six billion dollars. Innovative companies in the optical networking business have also been built on this model.

Blockbuster model

Many industries are dominated by 'blockbuster' strategies: strategies that depend on identifying and focusing on single products with massive market potential. This model is dominant in, for example, the movie, music, pharmaceutical and software industries. These industries are all characterized by increasing returns – huge development costs, but low marginal costs.

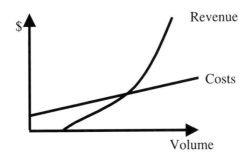

Figure 4.4 The blockbuster model.

Blockbusters require critical marketing mass. Once a potential blockbuster has been identified, then it needs to be marketed very strongly. This is particularly important if the product has an opportunity to benefit from network effects.

Profit multiplier model

The profit multiplier model (Figure 4.5) is based on reusing the same asset over and over. The classical model is a Hollywood company that develops a movie, then launches tape and DVD versions, CDs with the soundtrack, toys, T-shirts, games, books and computers games based on the same themes and characters. The model is also common in

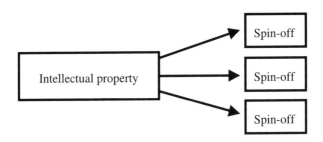

Figure 4.5 The profit multiplier model.

software companies, where the same code can be reused in many different applications. Some software companies even launch software that fails and then relaunch pretty much the same software in a new context, where it becomes hugely successful.

Never miss another moment

Sport television is a big and complex business. The rights to cover live sports events with video are traditionally sold by the relevant sports associations to sports marketing companies, which organize the video production, etc. The marketing companies will then recoup their investments as they resell local distribution rights to television companies around the world.

However, by 2000 it had become clear to a few people in the industry that television would not be the only valuable platform for these video experiences. There was a significant additional opportunity: to deliver short video clips of key events to new handheld devices and car entertainment systems. A new company called Worldzap was founded in year 2000 as a joint venture between Prisma Sports & Media, The Fantastic Corporation and ETF Group to pursue the enormous opportunity.

Delivery was critical. Sports events have to be covered live, or very close to live, in order to create the full excitement. This was exactly what Worldzap would do – it would deliver these video clips almost instantly after a goal, a record or another dramatic event had taken place. Fans that couldn't be by the screen as the action was on would still be kept in touch through this typical profit multiplier model.

Network-based Value Models

The network-based value models are based on the conscious effort to create a network that stimulates automatic escalation of the business. There are four main approaches:

- community-building model
- platform and standards model
- digital exchange model
- niche dominance model.

Community-building model

The concept of community building is perhaps the simplest of the network-based value models. It is based on the development of a network of users who communicate with the company and with each other, which enhances the value of the product for each user and increases the brand-building opportunity for the supplier. It can be one of the strongest value models in the new economy, since successful implementation can lead to an escalation of the business that continues to feed upon itself and which leaves competition further and further behind. It is based on one of the most significant pieces of nonlinearity in the new economy.

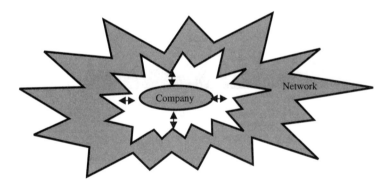

Figure 4.6 The community-building model.

It can be argued that few companies have had as big an opportunity to create network effect as Netscape. The company had a clear first-mover advantage, but it chose, erroneously, to pursue a business model based on charging for the software. What it could have done was to pursue a community-building model by using its status as a software provider to create the leading portal (as Yahoo! did) and invest extensively in its content partners to reap the financial benefits that it created for these partners by letting them participate in their portal. Licensing software is a linear model, whereas building communities is a model based on positive feedback loops.

Platform and standards model

This model is based on the creation of a position where the company supplies a solution that numerous other companies must support in order to be attractive to their own clients. A platform can be, for instance, a website with a portal status, a digital exchange, a software package or a widespread hardware unit. A successful platform is one that enjoys wide distribution and which provides the basic communication tools that other companies have to interface with.

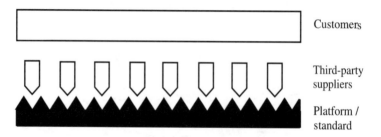

Customers

Third-party
suppliers

Platform /
standard

Figure 4.7 The platform and standards model.

Standards can be low level or higher level. Again, the key is that they are so dominant that many other companies need to interface to them in order to be attractive to their users.

The company that has positioned itself as a successful supplier of platforms or standards will enjoy some of the marketing power of any other company that delivers complementary products. The company will thereby enjoy an increasing stream of recurring revenues that have a limited relationship with its own marketing and development efforts. It piggybacks on the efforts of other companies to enjoy commercial leverage.

Providing the bridge

Mediagateway was launched in Denmark in 2000 to solve a problem in the Scandinavian e-commerce market. Scandinavia had many different digital television systems, but they were based on different digital platforms, and they were each too small to justify development of extensive e-commerce solutions. Media-gateway was thus founded to create one platform that could unite a number of e-commerce providers and bring their joint services to the digital e-commerce platforms.

Digital exchange model

A digital exchange is a special case of a platform and a network. It is a virtual marketplace where transactions take place through electronic interchange. A digital exchange provides value in numerous ways. The most important is to enable buyers and sellers to reach each other. But it can also create value by providing transparency, product standards, community and liquidity, among other things. The exchange provider will typically charge a modest fee per transaction.

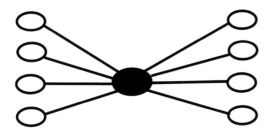

Figure 4.8   The digital exchange model.

Digital exchanges are efficient only if there is a critical mass of usage, and it is thus very difficult to build an exchange up to the point where it starts feeding on itself – typical investment levels required are

500 million or even one billion dollars. The market is ruthless – the largest exchange in a market space will often become completely dominant, while smaller exchanges will merge, be bought or fail.

Distant views

Remote-I was formed in 2000 as an example of a digital exchange based on a new and radical idea. With new GPRS and UMTS telephones, webcams and digital camcorders coming up everywhere, there would soon be hundreds of millions, if not billions, of devices in the world that could record digital video. Most people in the developed world would have at least one such device. The digital mass represented as video would thus explode, and this was the ideal situation for an electronic market place.

Remote-I's idea was to create a global exchange for video clips captured by anybody, anywhere. It could be anything from teenagers posting a video statement as a part of a dating service to private people filming news events on the spot, as they unfolded, and selling the clips to the highest bidding news agency. The exchange, would in any case connect people and provide them with a system to exchange and deliver their video clips.

Niche dominance model

It can be argued that the most successful businesses are, in effect, all based on some form of niche dominance. They are based on the assumption that a company needs to focus on a market area that is small enough and so well-defined that the company has a chance of dominating it. Such a niche can be of many different sizes, but many companies have avoided disasters by redefining their focus to niches that they were able to dominate. Only after they have achieved this position did they move on to add new niches.

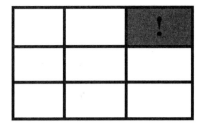

Figure 4.9   The niche dominance model.

Timing-based Value Models

Timing has always been a critical element in business, but it has become much more important in the environment that we call the 'economy of speed'. There exist two basic approaches for companies that
focus on time-based strategies. One is the ad hoc approach, which bases
the company on a specific first-mover advantage, e.g. a new innovation or a very aggressive marketing move. The other approach is the
'organized-for-speed' model, in which the company is organized in
such a way that it can move faster than its competitors *at all times*.

First-mover advantage model

The first-mover advantage model (Figure 4.10) utilizes elements of
increasing returns. The first mover can file for the best patents, define

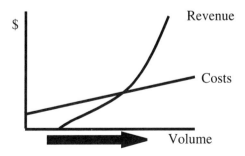

Figure 4.10   The first-mover advantage model.

the standards in its industry, attract the best people, create a network effect for its products, get its brand associated with the industry, etc.

The challenge for the first mover is not competition, but the development of its market. A first mover may often underestimate its own strength during the initial period where the market is small. There are many stories of new-economy companies and projects that enjoyed first-mover advantage but didn't manage to turn this into subsequent dominance, simply because they vastly underestimated how much the market would grow. Not only this, they also underestimated how long it would take before the market would take off and how demanding it would be to create an offering that appealed to mainstream customers. Generally, new markets get bigger, but take longer to get off the ground, than people expect.

Organized-for-speed model

A company that is organized for speed is organized to create first-mover advantages all the time. It analyses all its processes and organizes itself so that internal communication lines are short, external communication networks are vast, and all procedures are as close to fluent and real time as they can possibly get.

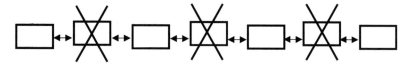

Figure 4.11   The organized-for-speed model.

Customer Relationship-based Value Models

There are four dominant value models, which are based on the special relationship between a company and its customers:

- secondary sales model
- mass customization model

- central aggregator model
- solutions provider model.

Secondary sales model

This traditional model is based on selling the core product with little or no profit, and then making the money on the add-on sales, e.g. selling razors very cheaply, but making the money on the razor blades, or giving away mobile phones to people that sign up for a multiyear carrier subscription.

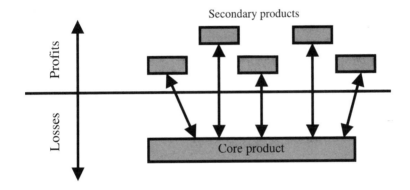

Figure 4.12 The secondary sales model.

The model can make considerable sense in markets with network effects, and it has thus grown in importance. Many Internet business models can, for instance, be described as secondary sales models. You make a website and give everybody free access to the information or entertainment it contains. The revenues come from secondary sales – e-commerce, sponsoring, advertising, etc.

Mass customization model

The mass customization model (Figure 4.13) is much newer. It combines customization with mass-market production. The pre-industrial economy was based entirely on customization: everything was manufactured to order. Ships, dresses and houses were all made to individual specifications. The industrial economy provided a change towards mass marketing. Standard designs were created for the entire market or for each of a number of marketing segments.

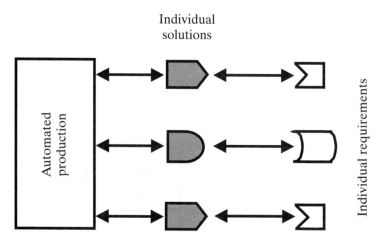

Figure 4.13 The mass customization model.

Mass customization combines the best of both worlds. It uses computer processing and digital communication to enable every customer to get their own design, but it uses technology to automate the manufacturing processes so that costs are kept low. Some of the key phenomena within the mass customization value model are:

- to involve the customer in the production and design process;
- to use computers and telecommunication technologies to manage the relationships from customer to suppliers;
- to enable the customer to generate more information about their own preferences through interaction with the company.

Alvin Toffler has written extensively about the first of these phenomena. He coined the term 'prosumer' to describe the consumer who contributes to the production process through collaboration with the supplier. This is not only efficient, but also emotionally evolving. It gives the satisfaction of having a stake in the creation and perhaps the possibility to follow the manufacture and delivery via the Internet. Alvert Bressand was describing the second phenomenon (use of computers for relationship management) when he coined the term 'R-Tech' for technology that facilitates relationship management (creating tailored products, recalling preferences, anticipating interests). John Hagel contributed to the understanding of the third phenomenon (enabling customers to learn about themselves) by suggesting that technology should not only help customers generate information about users, but also help users generate information about themselves.

A successful mass customization model creates strong customer preferences and loyalty, and is thus the basis for abnormal revenue growth and profit margins.

Central aggregator model

The central aggregator model (Figure 4.14) is characteristic of the digital age. An aggregator can be an Internet portal, a television network, or a digital exchange combining services from multiple partners.

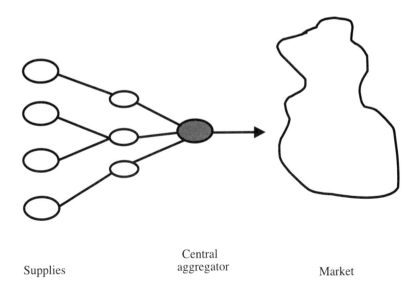

Supplies   Central aggregator   Market

Figure 4.14 The central aggregator model.

Digital communication based on open standards makes it possible for several companies to aggregate input from others into portals. A key attraction of this model is that the aggregator is in a position to attract more users than many of its partners, and yet it doesn't have to go through the expensive process of actually creating the original material. However, it is a business model that tends to create a few very big winners and several smaller 'me-toos'.

Solutions provider model

A solution is normally a combination of a product and a service that provides a broad benefit to the customer. The more complete the solution, the greater the benefit for the end user. The solutions approach has thus been pursued successfully by numerous companies in both the old and new economies. Many large companies in the IT industry saved their futures when they changed from a product approach to a solution approach. For example, a critical part of the turn-around of IBM in the late 1990s was when the company decided to sell products from other suppliers, as well as its own, which it knew the customer would be better off with.

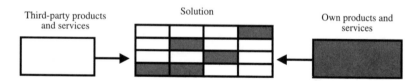

Figure 4.15   The solutions provider model.

Cost-based Value Models

There are three major cost-based value models, all of which were common in the old economy but can be exercised even more efficiently in the new economy:

- critical scaling model
- low-cost business design model
- consolidation play model.

Critical scaling model

The critical scaling model (Figure 4.16) is based on building the company to a size where its costs are considerably below the costs of its competitors. This model worked efficiently in several old-economy industries, where companies became highly profitable as they scaled up to the point where marginal returns reached zero. They achieved 'economies of scale'. However, it takes on a whole new meaning in business with increasing returns. The company that can scale fastest will create not only a temporary advantage but often a *permanent* advantage due to path dependency. The company that gets big fast will experience automatic escalation and will soon be able to out-price and out-gun its competitors on all fronts.

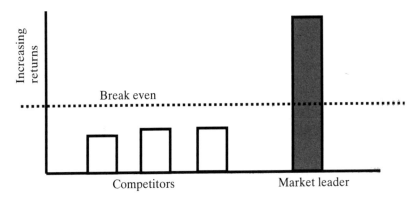

Figure 4.16   The critical scaling model.

Low-cost business design model

The essence of the low-cost business design model (Figure 4.17) is to automate processes that cost money in order to achieve lower overall costs. The preferred approach in the new economy is to automate any process that involves exchange of information. An efficient business enables this to take place between people, jellybeans and computers in a fluid, real-time flow.

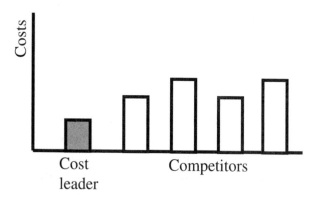

Figure 4.17 The low-cost business design model.

Consolidation play model

A consolidation play model (Figure 4.18) is the process by which a company merges with others to create leadership in its sector. While this role is often performed by the biggest players in the sector, experience from the new economy has shown that even complete outsiders such as AOL, Yahoo!, Worldcom, Pacific Century Group and Vodafone can enter an industry and create large consolidation plays within a relatively short span of years.

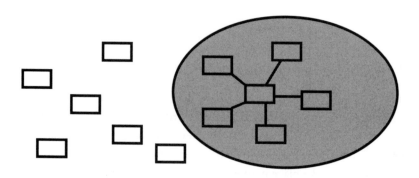

Figure 4.18 The consolidation play model.

CHOICE OF VALUE MODELS

The different value models for the new economy are not mutually exclusive. Most successful companies are simultaneously pursuing several of these models in a mix of strategies that seek to benefit from combinations of network effects, minimal marginal costs, path dependency and leverage. The choice is a question of understanding the opportunities in the market and the way the company can pursue them. It is a question of marketing.

THE DIGITAL SCHOOL

'Take a moment out of the heat of your current pitched battle and chew on the implications of this thought. We are right now in the very early stages of a new economy, one whose core is as fundamentally different from its predecessor as, say, the automobile age was from the agricultural era. If you grasp this premise it's much easier to understand a lot of what's going on around you, including why a seemingly unrelenting tsunami of change keeps washing over you and your business'.

John Huey

CHAPTER 5

THE DIGITAL SCHOOL

*T*he digital school of marketing, as we have chosen to call it, is based on all the observations that we have seen so far. It is based on the recent developments of digital data processing technology, the compression of time, and the new opportunities to find value.

So, how do we describe the substance and practicalities of this school? One way would be to approach it in the framework of the way companies do their *marketing planning*. We cannot describe this by focusing only on what is new and ignoring what remains from the previous schools. This would be like trying to speak English by using only the expressions that have evolved over the last 20 years.

THE CORPORATE PLANNING CYCLE

It is sometimes stated that the new economy is so complex that we cannot plan it. This is not true. Any successful new-economy company has to make elaborate marketing planning decisions. It just does it much faster than in traditional companies. It is rarely an annual plan; plans are made ad hoc, all the time, often to cover specific aspects and, from time to time, to cover the whole. This happens, for instance, when a start-up company is looking for financing, and before it goes public. Even if the company doesn't require financing, it will still need to plan, because it is only through detailed planning that we get to the essence of what we want to achieve, and how we will get there. This is called 'the planning cycle'.

The corporate planning cycle (Figure 5.1) involves a number of steps that are pretty much the same in any company, new economy or not:

1. Think very hard about the company and its marketing. If this thought process is done in a reasonably structured way, then it is referred to as a 'marketing audit'.
2. Write the long-term marketing plan. The meaning of 'long term' will depend on the company and its market.

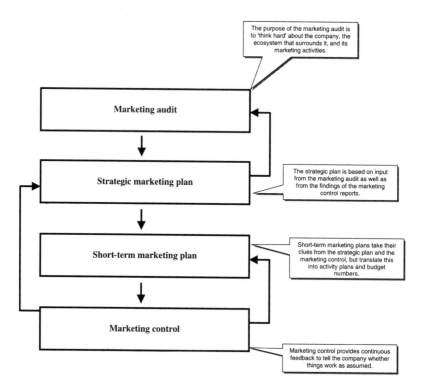

Figure 5.1 The corporate planning cycle. Planning is more erratic in the new economy than in the old, but the basic cycle structure is the same.

3. The long-term plan creates the basis for short-term plans.
4. Short-term plans needs regular follow-ups though marketing control.

Most professional companies work like this. Nothing has changed, except the intervals and perhaps the regularity (a very volatile environment can call for more frequent ad hoc rewrites).

Companies make other plans as well, of course. They may, for instance, make release plans, launch plans, production plans, communication plans or development plans. But, for the purposes of explaining the marketing challenge in the new economy, we will stick to these four key exercises (anyone who can master these should be in good shape to develop other marketing-oriented plans). Before we describe each of the four steps as practised in the new economy, we must look briefly at how companies gather the necessary information: 'marketing research'.

MARKETING RESEARCH

A basis for all marketing planning is marketing research which comprises the methods a company can use to collect information about what is happening in the market and the effects of its own marketing efforts. The traditional marketing schools have divided marketing research methods into two main categories. *Desk research* is based on collecting data that other people have generated. It can, in most cases, be done without leaving the office, hence the term 'desk'. The main sources of desk research are statistical publications, press scanning services, periodicals, multiclient studies and online databases. Searching the Internet is also a form of desk research. Figure 5.2 shows how information flows when a company performs desk research.

Field research (Figure 5.3) is more cumbersome. Researchers have to leave the office to collect data in the field. There are many different field research methods, but the most important ones are:

- focus groups
- informal interviews

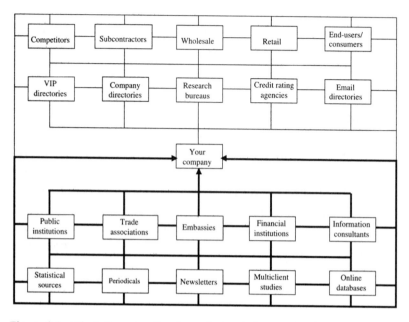

Figure 5.2 The information flow in marketing desk research. The bold lines high-light how information flows as companies conduct traditional desk research. Desk research draws on statistical resources, periodicals, newsletters, multiclient studies and online databases to gather information. The data are collected directly (typically via the Internet) or via intermediaries, such as trade associations.

- key interviews
- distribution research
- retail panels
- radar
- supply mapping
- consumer panels
- concept tests
- product tests
- advertisement pretests
- packaging pretests
- concept/product tests
- market maps.

Appendix C gives a short overview of these research methods.

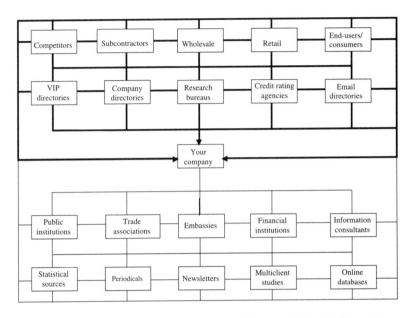

Figure 5.3 The information flow in marketing field research. The bold lines indicate flow of information as companies conduct traditional field research. Field research is based on direct collection of data from competitors, subcontractors, wholesale, retail and end-users/consumers. The company can collect the data directly or it can use intermediaries, such as research bureaus and directories.

The terms 'desk research' and 'field research' would cover all corporate market research in the traditional marketing literature. However, if we look at how new-economy companies operate, we often find that the bulk of their market information comes from a third category of market research: *on-line research* (Figure 5.4). This utilizes online connections between the company and its partners and customers. Online market research can be done in many different ways:

- *Click-stream and page-view tracking.* The company uses real-time analysis of the customers' behaviour on the website to map how they move from item to item, what they buy, etc. Several companies provide great software packages to facilitate this exercise.
- *Link tracking.* This tracks how many and what kinds of websites are providing links to your own site. There are software packages that automate this function. The benefit is that it teaches the

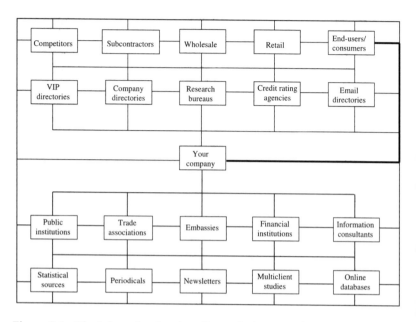

Figure 5.4 The information flow in online marketing research. The bold line highlights information flow as companies conduct online marketing research. This approach is unique as it connects the company directly to its customers on a one-to-one basis and in real time. The approach is complementary to the traditional desk and field research approaches, but it is unique in that it provides extremely high quality of data with a minimum of time delay and often at very low cost.

company about what kind of companies and people exist in the commercial ecosystem relating to the company.

- *Email paging.* Mihaly Czikszentmihalyi, Professor of Psychology at the University of Chicago, pioneered this method. It works by giving a number of people pagers that alert them at random moments during the day. Each of them then writes a sentence or two describing what they are thinking and doing when the pager alerts them.

- *Email interviews.* Many companies use email to gather instant feedback on product and marketing ideas. They can use information about their customer database to reach all customers within short timespans.

- *Online votes.* The company asks visitors to its website which of several alternative offerings they would prefer.

- *Wish lists and shopping baskets*. The company allows customers to add products that they might buy to an electronic wish list or shopping basket. This creates information about likely future sales.
- *Cyber-panels*. People volunteer information about their habits and preferences.
- *Virtual pretesting*. The company presents descriptions of potential products, adverts and designs on its website to see how many people click on them to gather more information. The company launches the product only if interest seems high.
- *Focus polls*. Live polls run simultaneously with a focus chat session in another window. The company can use the results from the poll as springboards for discussion.
- *One-to-one marketing*. The ultimate way to gather information about the market: obtain direct, real-time information about what each individual really wants.

These new marketing research methods enable the company to generate extremely detailed information in real time and on a one-to-one basis; they are essentially changing the core paradigms of marketing research, especially in relation to collection of information about the company's own planned and actual marketing activities.

LONG-TERM MARKETING PLANNING IN THE NEW ECONOMY

Imagine you are running a dynamic company in the digital economy. You have to write a strategic marketing plan. So what do you do? Do you make a spreadsheet with a lot of numbers plus half a page of text? Or do you sit down with some of your team and ask some basic questions first? Successful companies are most likely to do the latter. They ask question such as:

- What are we essentially trying to achieve?
- Are we in the right value space?

- Do we really understand the market?
- Are we approaching the market in the right ways?
- Are our speed and timing right?

If this discussion is just remotely systematic, then you are embarking on a 'marketing audit'.

THE MARKETING AUDIT

Marketing audits were practised regularly by the consulting company Booz-Allen-Hamilton as early as 1952, but structured, public descriptions of the process did not appear until 1959, when the American Management Association published a number of papers about the subject. The marketing audit is a critical review, whereby the marketing approach of the company is examined in its entirety. It is a parallel to the legal audit (review of a company's legal structure and commitment) or the financial audit. It may be divided into 11 separate steps, as shown in Table 5.1.

Table 5.1   The 11 steps of a marketing audit

Question	Corresponding audit
What are we essentially trying to achieve?	1. Defining corporate mission statement
Are we in the right value space?	2. Value audit
Do we really understand the market?	3. Economic environment audit 4. Business ecosystem audit 5. Competitor audit 6. End-user audit
Are we approaching the market in the right ways?	7. Product positioning audit 8. Pricing strategy audit 9. Distribution strategy audit 10. Promotion strategy audit
Are our speed and timing right?	11. Speed and timing audit

The process is cyclical: the 11 audit steps are performed in order, then revisited if necessary. The result of the audit may be that the company needs to be fine-tuned here and there, but it could also be that it needs to totally reconsider the way it is managed or even change its basic mission statement.

Audit Step 1: Defining Corporate Mission Statements

The first step in the audit is the most important one. It is to define the company mission statement. It will typically define a market area that the company aims to lead. This may, for instance, be:

- a demand that the company is particularly good at satisfying
- a customer segment
- a product segment
- an interactive user community that it will serve
- a one-to-one solutions approach.

The definition may also include:

- an online connectivity concept (such as a mobile phone)
- a demographic segment
- a pricing segment
- an image
- a benefit.

It may also refer to different value strategies:

- brand-building model
- disruptive invention model
- build-to-be-bought model
- blockbuster model
- standards model
- community-building model
- platform and standards model
- digital exchange model
- profit multiplier model

- first-mover advantage model
- organized-for-speed model
- secondary sales model
- mass customization model
- central aggregator model
- critical scaling model
- low-cost business design model
- consolidation play model.

The stated objectives may also go beyond what the company wants to achieve, by describing critical qualitative criteria for achieving it, such as key competencies, key staff, production resources or experience.

Thousands of books and articles have covered the theory and philosophy of corporate mission statements. A few have become classics, such an article written by John McKitterick in 1957, which included the first clear description of the modern marketing concept: that companies aim to *identify and satisfy the demand of consumers*. Another classic article, 'Marketing myopia', written by Theodore Levitt in 1960, explained how companies and entire sectors could lose their edge because they erroneously defined their mission as *to deliver a given product* rather than to *satisfy a given demand*. Levitt wrote:

> The railroads did not stop growing because the need for passenger and freight transportation declined. That grew. The railroads are in trouble today not because the need was filled by others (cars, trucks, airplanes, even telephones), but because it was *not* filled by the railroads themselves.

The railroads had, he concluded, defined their purpose in the wrong way. Their *raison d'être* was not to move trains around, but to move people and freight around. Many companies in other industries had committed the same error, Levitt claimed, for example the movie studios:

> Today TV is a bigger business than the old narrowly defined movie business ever was. Had Hollywood been customer-oriented (providing entertainment), rather than product-oriented (making movies), would it have gone through the fiscal purgatory that it did?

Another important contributor to the thought process was Peter Drucker, who is known for his 1964 article, 'The big power of little ideas,' in which he pointed out that a mission statement had to be precise and specific in order to be effective. Mission statements are often incomplete or plainly misunderstood. Some of the most common errors are:

- *Confusing operational goals with missions.* Sometimes mission statements define how much the company should earn or how large a market share it should achieve. These may be relevant objectives at the operational level of a department or a salesperson, but they do not convey the message needed on a strategic level as they do not indicate how the quantitative target should be met. They are not suitable as corporate mission statements.
- *Confusing strategies with missions.* The mission statement should be lasting, whereas strategies for fulfilling it are constantly evolving with the market and the company's position in it.
- *Confusing the mission of the company with the missions at its business levels.* What is done at business levels is merely a strategy for fulfilling the corporate mission.
- *Defining the mission too narrowly.* Keep in mind that the market segment in which the company is operating might be served in completely new ways in the future. Maybe a post office's letter mail department defines its role not as delivering letters, but as delivering messages. If so, then perhaps they should play a role in electronic communication? This is why corporate missions are often defined by the customer demands the company aims to fulfil.

Product-oriented approaches to corporate mission statements becomes more and more problematic as every company increasingly depends on strong ties with other companies to develop the critical network effect. What a company delivers may matter less than what network it has created. The network is the business.

Audit Step 2: Value Audit

Defining how to make the money is the next step in the audit. We have already been through the typical value strategies in the new economy. So which ones do the company pursue? It will typically be several. The most useful way to view the choice of value strategies is to evaluate them in the perspective of the four leading characteristics of the new economy: how the strategy can take advantage of network effects, minimal marginal costs, path dependency and leverage.

The following tables show how each of the value models typically takes advantage of the key characteristics of the digital economy (where ★★★ means maximum utilization). The *asset-based* value models are based on different utilization of the characteristics of the new economy (Table 5.2).

Table 5.2 Asset-based value strategies and their benefits from four characteristics of the new economy

	Network effects	Minimal marginal costs	Path dependency	Leverage
Brand-building model				★★★
Disruptive invention model			★★★	
Build-to-be-bought model			★★★	
Blockbuster model	★★	★★★		
Standards model	★★★		★★★	★★★

The *network-based* models are, of course, based primarily on taking advantage of network effects, but most of them are also leaning against other opportunities in the digital economy (Table 5.3).

Table 5.3 Network-based value strategies and their benefits from four characteristics of the new economy

	Network effects	Minimal marginal costs	Path dependency	Leverage
Community-building model	★★★	★★	★★★	★
Platform and standards model	★★★	★★	★★★	★★
Digital exchange model	★★★	★★	★★★	★
Profit multiplier model	★★	★★★		★★★

The *timing-based* models are, more than anything, about getting to a leading position before the competition. They are therefore based mainly on benefiting from path dependency. However, moving quickly will typically also mean aggressive use of leverage and creation of strong network effects (Table 5.4).

Table 5.4   Timing-based value strategies and their benefits from four characteristics of the new economy

	Network effects	Minimal marginal costs	Path dependency	Leverage
First-mover advantage model	★★★	★	★★★	★★★
Organized-for-speed model	★★	★	★★★	★★

The *customer relationship-based* models are based primarily on network effects, minimal marginal costs and path dependency (Table 5.5).

Table 5.5   Customer relationship-based value strategies and their benefits from four characteristics of the new economy

	Network effects	Minimal marginal costs	Path dependency	Leverage
Secondary sales model	★★	★★	★★★	
Mass customization model	★★★	★★	★★★	
Central aggregator model	★★★	★★★	★★	★★

The *cost-based* models are focused primarily on utilizing the minimal marginal costs that many new economy companies can enjoy (Table 5.6).

The acid test for the company is to consider whether the models it is pursuing will actually enable it to benefit in any major way from any of the key characteristics of the digital economy.

Table 5.6 Cost-based value strategies and their benefits from four characteristics of the new economy

	Network effects	Minimal marginal costs	Path dependency	Leverage
Critical scaling model	★★★	★★★	★★	★★★
Low-cost business design model		★★★		
Consolidation play model	★★★	★★★		★★★

Audit Step 3: Economic Environment Audit

The next step in the marketing audit is to evaluate the overall economic environment in each of the market areas the company operates within. Relevant variables include:

- economic climate
  - business cycles
  - interest rates
  - inflation
  - availability of qualified work force and other resources
  - market segment's purchasing power
- political environment
  - monopoly law
  - price controls
  - environment legislation
  - tax laws
  - duty laws
  - personnel legislation
- demographic conditions
  - age group distribution
  - income levels and distribution
  - cultural patterns
  - employment patterns

- lifestyle trends
  - new lifestyles
  - new attitudes to the product
  - new purchasing habits
  - new usage habits

Audit Step 4: Business Ecosystem Audit

Digital business obviously cannot thrive in isolation. Take, for instance, software. Any software from any vendor must interface with other software from numerous other vendors. It is hard to think of an exception. The most successful software is often that which frequently interfaces with the highest number of other software packages. Highly successful software vendors have thousands of other companies writing software applications for their platforms, just as successful websites are linked to many others. This means that companies in the digital economy operate with a very large number of partners. But how do we describe how these partners work together?

Value chains

One of the most commonly used terms in marketing is 'value chain'. A value chain is a description of value-added steps from basic resources (such as raw materials) to end-user proposition (the product and service that the end user experiences). The value chain follows the products. It shows how a product or service is developed and sold as different players add value in a step-by-step approach. It also follows the money, if you read it in reverse. The product or service is built from left to right, and money flows from right to left. The value chain, in its simplest form, is shown in Figure 5.5.

Value chain diagrams could, of course, be much more complex. They may, for instance, include primary production, aggregation into a more sophisticated product, sales, physical distribution, logistics, service and support. They may also be reversed, beginning with customer demand, and moving backwards to describe how this demand is filled (Figure 5.6).

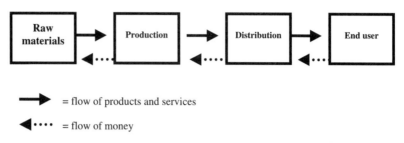

➡ = flow of products and services

◄•••• = flow of money

Figure 5.5 Traditional illustration of the value chain.

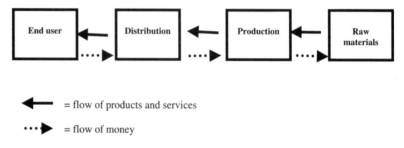

◄— = flow of products and services

•••► = flow of money

Figure 5.6 Demand-oriented illustration of the value chain.

Michael Porter, a university professor from Harvard University, popularized the value chain concept in 1980, with his book, *Competitive Strategy: Techniques for Analyzing Industries and Competitors*. Porter used the value chain as a reference model for a series of instructive clarifications of the desired roles of the company in the marketplace. He described the company's relationship with competitors, suppliers, distributors and alliance partners better than any marketing author had done before. He pointed out that some types of competitors can actually help a company: 'While competitors can surely be threats, the right competitors can strengthen rather than weaken a firm's competitive position in many industries.'

Why? For instance, because: 'competitors can absorb fluctuations in demand brought on by cyclically, seasonally or other causes, allowing a company to utilize its capacity more fully over time.'

There were more reasons too:

Competitors can enhance a firm's ability to differentiate itself by serving as a standard of comparison . . . A high-cost competitor can sometimes provide a cost umbrella that boosts the profitability of a low-cost firm. It is a common view that industry leaders provide a price umbrella for industry followers, and this is indeed the case in some industries . . . Having competitors can greatly facilitate bargaining with labor and government regulators, where negotiations are partly or wholly industry wide . . . A role of competitors that is hard to overestimate is that of a motivator. A viable competitor can be an important motivating force for reducing cost, improving products, and keeping up with technological change.

This doesn't mean that all competitors are beneficial, of course. Some are beneficial, some are not; the key observation is that smart companies to some degree can choose who they want to have as their competitors. They can then make different alliances with the rest of the suppliers in their space.

Porter made another vital observation. He noted that a key consideration should be which elements in the chain the company wants to cover by itself. The more elements the company covers, the more control it will have. However, covering many steps in the chain may exclude the company from making alliances with other companies that could perhaps cover some of these elements better. Vertical integration could be a hindrance to horizontal and virtual integration and to the development of network effects.

Most of Porter's models belonged to the traditional institutional school of marketing, since they dealt with the marketing organization process. His contributions to marketing thoughts make him the recognized leader of this school, and probably one of the top 10 thinkers in all marketing theory. However, despite his models, it didn't take long before an alternative came up. Some began to talk about 'e-chains' and 'value webs' instead of value chains.

E-chains and value webs

The futurist, Paul Saffo, first used the term 'value web' to describe the multiple interactions needed to survive in the new economy, in his article, 'Disinter-remediation: the surprising impact of information systems on markets and organizations':

> The notion of a value chain is as much an optical illusion as the notion of disintermediation. One doesn't have a single-transaction relationship with a seller, but

rather, the relationship arises out of multiple interactions. And these quickly evolve into a web of interactions, not a chain. Some interactions are immediate and direct, while others are highly indirect, transmitted through intermediaries serving the interests of both buyer and seller.

The term 'e-chain' was based on similar thinking. One of the first to use it was Michael Hentschel, Managing Director of TechVest International, who described it in the market research report, *Portals to Profit: E-Commerce Business Models and Enabling Technologies* (1999). Here is how he explained the idea in an interview with *Washington Technology* on 7 February 2000:

> E-chains are my term for the B-to-B [business-to-business] phenomenon. The B-to-B are mostly vertical Web sites but can also be horizontal. A B-to-B is a Web site on almost any vertical level now, whether it's steel, other commodities or roofing.
>
> Business-to-business types of what were once called extranets are now turning into portals that are a combination of very specific hubs that deal in any information that has to do with a certain industry.
>
> The first person who puts together that B-to-B network is putting together various parts of an e-chain, an electronic chain of linkages that used to be called a network or intranet. If you put together the proper pieces of the e-chain and make them sufficiently open to the outside world to be used by a lot of people in that particular industry, people will gravitate to it, because it will be the easiest and fastest way to do transactions.
>
> It's not really a matter of cost; the cost is already there. If you make something more efficient, you will save people money, and they will flock to use those parts of the e-chain.

The concepts of value webs and e-chains were reactions to the deficiency of the value chain models, which to some appeared to have been inspired by the assembly lines in factories of the industrial society. Raw material comes in through one door, cars come out through the other; value is added in a series of successive steps to get from raw material to car. However, the typical processes in a digital economy are rather more complex. There is information flowing constantly from the producer to the consumer, but also from the consumer to the producer, and between consumers sharing knowledge about the products. The same goes for digital money flows: they can go in both directions. Just think about Ebay, the digital bazaar where Web surfers trade with each other over an electronic platform, or Nasdaq. So, the best way to describe the commercial framework is to follow the flow of bits and bytes, whether they represent information, money transfers or transfers of digital products.

However, the e-chain and value web concepts didn't really fly in the boardrooms because:

- the market interaction in a digital market doesn't look like a chain when you draw it – it's more complicated;
- the relevant relationship goes beyond what can be represented by flows of bits and bytes: new-economy companies are, for instance, very often investing actively in the companies they do business with.

The solution was the 'ecosystem' approach, which (apart from perhaps having a more hip name) had the advantage of describing more aspects of interaction in a complex environment.

Ecosystems

James Moore first used the term 'ecosystems' in a business-related sense in his article, 'The advent of business ecosystems' (*Upside Today*, 1995). He began his article with a description of some characteristics of ecosystems in nature. Then he compared these to the commercial and financial systems surrounding large companies in the computer industry. These companies not only traded with each other, he noted; they also shared ideas, formed consortia, stimulated user communities, invested in their partners, and much more. He concluded:

> Over the past few years we have struggled to find the right description for close communities of organizations – usually taking guidance from one or more powerful lead firms – that are not simply managing themselves but are transforming the economic worlds in which they live. My own preferred term is 'business ecosystems'.

Figure 5.7 shows an example of a business ecosystem, where the arrows illustrate sales of products and services. However, other arrows could be added to show information relationships, development and production relationships, marketing relationships and financial relationships.

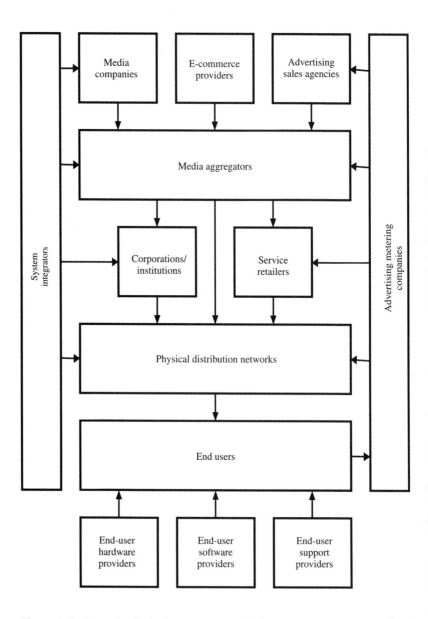

Figure 5.7 Example of a business ecosystem. Business ecosystems are networks of companies that are linked electronically, financially or commercially. The diagram shows a simplified example of a business ecosystem, in this case for the digital broadcasting market.

If Figure 5.7 looks a little bit complicated, then it's because it is inaccurate: a realistic diagram would be *very* complicated. The TMT sector is now serviced by companies that specialize in selling diagrams with overviews of how companies in their sectors are related. The diagrams are complex and three-dimensional. There are so many tiny lines between each of the players that some people can't see them. And yet, an insider in the industry will know that even the most complex of these diagrams will describe only a tiny fraction of the movers and shakers in these industries. Furthermore, they describe only a tiny fraction of all the bonding lines between the different players. What's more, these bonding lines change all the time.

The commercial ecosystem in the digital economy displays constant change. TMT companies behave rather like insects in the Amazon jungle. They move around constantly, grow, eat each other, mate, fight, work together, die. So, when we describe a commercial ecosystem, we must consider many different angles of the cooperation (or battle). Brian Arthur described the role of these relationships: 'Players compete not by locking in a product on their own but by building webs – loose alliances of companies organized around a mini-ecology that amplifies positive feedbacks to the base technology.'

One of the areas in which the ecosystem mentality has been crucial is in high-tech development environments. Development facilities have often been called 'laboratories'. There is some sort of magic aura surrounding the word. It was in such laboratories that companies like IBM, AT&T and Xerox made their fantastic breakthrough innovations during the 1960s and 1970s. The approach with development laboratories was to isolate excellent research people completely from daily operations and commercial life so that they could focus entirely on development of new stuff. This approach was largely abandoned in the 1990s, because, as the markets opened up and open standards became the norm, it became clear that successful innovations could no longer be shielded from the rest of the world: companies missed out on the network effect if they didn't connect with the world. Today, it is more and more common to bring suppliers, customers, alliance partners and investment partners into the development process.

The same applies to production and marketing. None of the processes involved in production and marketing can be done profitably if they are not executed as a collaborative process involving numerous companies, even if this means (as it frequently does) that there will be many to share the rewards. Some of the world's most successful software companies are, for instance, known for giving away up to 80% of their revenues to their commercial partners.

The concept of ecosystems is an example of an idea described in previous marketing schools (in this case the institutional school) that has been rewritten by authors of the digital school to reflect the new realities.

Evaluating the company's strengths within its ecosystem

Now that we have addressed the way networks of companies in the new economy are described, we should move on to look at how a company can evaluate its own strength within such a network. The first approach here is to draw up the commercial ecosystem, including all the important players and categories of players. In this part of the audit, we study how attractive the business sector the company operates in is (or isn't) and how the company may best profit within it. Factors that may be examined include:

- attractiveness of the sector in the future
  - technical standards
  - patents
  - possibility to develop network effects
  - switching costs and other barriers against new suppliers
  - cost trends
  - demand factors
  - growth
  - importance of dominance
  - importance of key staff and other critical resources
  - legal aspects
  - price trends
  - profit margins

- size
- technological developments
- the company's competitive strength in the sector
  - first-move advantage
  - cost level
  - critical deals, alliances and contracts
  - network effect
  - customer loyalty
  - distribution network
  - dominance
  - flexibility
  - geographical location
  - growth
  - image position
  - key staff
  - marketing
  - organization
  - patents
  - profit margins
  - salesforce and sales methods
  - size
  - technological position
  - leverage

The key purpose here is to identify qualitative aspects of importance. Maybe there is a new technology evolving that may turn everything upside down. Maybe a key competitor has gone public and raised cash before your company. Maybe a company is sitting on a critical patent. Maybe a company has developed a network effect that makes it virtually impossible to beat. Maybe a company is offering elements of its products for free, while you are still charging for the equivalent.

An analysis of the attractiveness of the sector and the company's position in it can provide the basis for strategic decisions regarding whether or not the company should invest/expand or harvest/terminate any given activity. Figure 5.8, which was originally developed by General Electric, illustrates how we can approach this problem.

Attractiveness of sector

		High	Medium	Low
Company's competitive strength in the sector	High	Expand	Expand	Expand selectively
	Medium	Expand	Expand selectively	Harvest/terminate
	Low	Expand selectively	Harvest/terminate	Harvest/terminate

Figure 5.8 Criteria for determining market commitment.

The key measures of competitive strengths in the old economy were capital resources and market share. In the new economy, it is more likely to be smart people, the ability to benefit from network effects and minimal marginal costs, and first-mover advantages and the use of leverage.

Building on strength in the ecosystem

Let us assume that a company concludes that its position in the eco-system is so strong that it should expand or invest. What are the practical options? Developing new products and services? Penetrating new markets? Selling more? Should the company buy licences, or sell them? Should the company participate in standards development, or buy into other companies that are already in the space?

This depends on many factors, but the most important is the company's own position – how well does it understand the market and how much of the relevant know-how does it possess? Figure 5.9 shows the typical approaches, depending on these considerations.

One of the typical new-economy issues here is the attitude to standards. If we look in the top left corner of Figure 5.9, we see that a common approach of a very strong company is to try to enforce its own proprietary standards. Slightly weaker companies will participate

Market power	Technology power		
	High	Medium	Low
High	• Development of proprietary standards • Internal development • Market-oriented acquisition	• Internal development • Market-oriented acquisition • Acquire licence • Participation in open standards development	• Joint venture • Marketing alliance • Acquire licence
Medium	• Internal development • Market-oriented acquisition • Joint venture • Participation in open standards development	• Internal venture • Market-oriented acquisition • Acquire licence • Participation in open standards development	• Know-how-oriented acquisition • Venture capital investment
Low	• Joint venture • Marketing alliance • Sell licence	• Venture capital investment • Know-how-oriented acquisition	• Know-how-oriented acquisition • Venture capital investment

Figure 5.9 Criteria for selecting strategic expansion models. The choice of expansion approaches depends largely on the company's strength.

in development of open standards. It should be stressed here that setting standards is a very slow process (sometimes almost unbearably so), and that any company depending on them has to develop a standards strategy for each important area, i.e. choose whether to support open or proprietary standards, and whether to lead or follow.

Table 5.7 shows the four approaches to market standards. There is no ideal approach: each strategy has its own strengths and weaknesses:

- *Participation in the development of open standards.* The advantages of this approach are that open standards are most likely to succeed, and that the company, through its participation in the definition

Table 5.7 Four approaches to market standards

	Open standards	Closed standards
Lead	Participate in the development of open standards	Launch proprietary standards either on its own or within an alliance of companies
Follow	Adapt to open standards	Adapt and in some cases pay licence

process, can influence and track the development. A company may also ensure that its own proprietary technologies become embedded into these standards. Open standards shift competition to branding, distribution, marketing, features and quality, and are thus beneficial for those who are fit for competition. It should be noted here that participation in an open standards forum, and what is said under negotiations, can bring real responsibilities. The company will, for instance, be required to disclose which relevant patents it owns and has filed for, and will have to agree to licence these on fair terms. In the USA, the participants in standard setting bodies are specifically required to license any essential or blocking patents on 'fair, reasonable and non-discriminatory terms'. One important issue of participation in these processes is the related resource requirements. Smaller companies can find these excessive, but they may choose to pool their efforts so that a number of them are represented jointly through one individual.

- *Launching proprietary standards.* The key advantages to this approach are that it is fast, the company can charge whatever royalties the market can bear, and it secures control over the future development of the standard. The best example in this is probably Sun Microsystems' launch of the programming language, Java, in 1995. When Java first came out, people doubted that Sun Microsystems could drive this as a proprietary standard, but it soon became clear, even to Sun's largest competitors (Microsoft, HP and IBM), that there was no way around Java. So, Sun succeeded in driving its own standard and even getting support from its closest competitors, who had no choice but to follow. The proprietary standards approach is typically (but not always) organized through a consortium of players, where each can bring assets, such as patents, branding, installed base and manufacturing skills to the table. The key disadvantage of the approach is that many such attempts, even when launched by impressive consortia of companies, eventually fail as open standards emerge. Furthermore, while increased size of a consortium increases the probability of success, it also slows down and complicates the definition process.

- *Adapting to standards.* This approach makes sense if the standards are not vital to the company or if it is not in a position to allocate the necessary resources or bring sufficient bargaining power to the negotiations.

Standards battles are long and sometimes dull, but they can be interesting, with secret side deals and knives in the back. An example is that many companies participating in standards initiatives are, in fact, hoping that these will not succeed – they hope that their own proprietary technologies will dominate by the end of the day. Another issue is that subgroups of participating companies will often engage in 'log-rolling agreements' – side deals with voting agreements between some of the committee members. Such agreements may specify that the voting consortium members are free of royalties while anyone else will have to pay. There may also be flanking manoeuvres if negotiations get too slow. Consortium members can thus launch, or appear to launch, proprietary solutions in order to shake up the talks. And there may be participants that pretend to support open standards, but who go on to develop proprietary extensions in order to steal the standards.

The approach towards standards is a key consideration for a company that wants to expand in the ecosystem. Another consideration is whether to *buy into the market* position or use *internal development*. Internal development is the path that market leaders often choose. The advantages are that it is motivating for the (development) staff, the development process can be fully controlled, and your company won't have to depend too much on outsiders. The key disadvantages are that it is slow and the company may not be able to allocate or find the needed resources.

An alternative to this approach is the *internal venture*, which creates a new, separate unit to handle the task. This ensures larger managerial freedom (at least in theory) and makes it easier to hire external staff without offending the insiders. It can also make it easier to remunerate differently, for instance with stock options based on the expectation of a potential initial public offering (IPO) of the separate unit if it succeeds. The challenge is the possibility of conflict with the rest of the company.

An alternative to internal development/venture is to enter into some form of direct commercial cooperation with other entities in the ecosystem. Common alliances include:

- production alliances
- selling licences
- acquiring licences
- market-oriented acquisitions
- know-how-oriented acquisitions
- joint ventures
- venture capital investments.

Production alliances include sharing logistic facilities (for shipping, billing, storage, computing, etc.), establishing joint production facilities, or engaging in other exchange of resources or know-how. These approaches can save time and money, but they can also involve coordination problems.

If the company has developed a key concept or technology, but doesn't have the market strength to build the network effects or benefit from first-mover advantages and minimal marginal costs, then it might make sense to *sell licences*. This is a fast approach that involves very low risks. However, the disadvantage is that the licence taker may block the distribution of the solution if it lacks market power or doesn't give the technology the appropriate attention.

The opposite strategy is to *acquire licences* from other companies if they have an obvious lead in the space. The advantages of this strategy are that it saves time, it is inexpensive (in the short run), and it may secure development of a market position that could otherwise be difficult to obtain. The disadvantages are the dependency that it involves, the limited development of know-how that it generates, and the risk that the licence provider may lose its competitive power. Plus the licence fees, of course.

Cooperation with other companies in the ecosystem can take stronger forms than simple licence agreements. It can involve financial engagement through either acquisitions, establishment of joint ventures, or venture capital investments. *Acquisitions* are often divided into 'market-oriented acquisitions' and 'know-how-oriented acquisitions'. The first regards purchase of customers, brands, networks and revenues. The second is to do with access to people, patents, standards and technology. Both can secure fast market entry, but they can also generate numerous problems as key people perhaps walk out or can't (or won't) figure out how to work together.

Joint ventures are the foundation of new units/companies in co-operation with partners. These are typically created for one of three reasons: securing a network effect, sharing costs, or complementing strengths. A joint venture to secure network effects is typically made between a group of companies that realize they are about to enter into a costly and risky war of standards to battle for the development of the leading network. To avoid this, they set up a joint venture company based on a compromise of their interests. The cost-sharing motive is simpler. Since the new economy is characterized by very large start-up costs and very low marginal costs, it can be beneficial for several companies to work together to share the start-up costs so that they are able to overcome the huge start-up barriers. Each of these companies might not be able to reach that point on its own. The third motive is about complementarity. A typical situation is what is sometimes described as the 'Siliwood syndrome' – Silicon Valley companies working with Hollywood companies. Hollywood produces digital entertainment, but such entertainment cannot be distributed in a world of ubiquitous computing without involvement of a heavy dose of technology, which is where the technology companies and their digital distribution solutions enter the scene. When content is king, distribution becomes King Kong. The key disadvantage of joint ventures is that there can be conflicting interests between the shareholders, and the shareholders may feel that they give away some of their potential upside to a partner.

The third main category of financial cooperation with other companies is *venture capital investment*, in which a company invests in a minority position in a smaller company. This may lead to an enhanced understanding of the technology and the market without requiring involvement in any of the day-to-day responsibilities of acquisitions. The initial cost is typically limited, but can be a precurser to a later-stage all-out acquisition. However, until such a step is taken, the control is very limited.

The final category of cooperation is *marketing alliances*. It is impossible to thrive in the new economy without entering into alliances. A company cannot reach the point where it enjoys a network effect with its customers without also creating a close network with other suppliers in the market. The typical marketing alliances in the new economy are:

- cross-sales
- joint promotion
- joint pricing
- integrated sales
- joint promotion of an organization
- joint promotion of facilities/staff
- joint branding
- sharing of sales administration.

These alliances are explained further in Appendix A.

The new economy is arguably more about cooperation and less about competition than the old economy. The different approaches to cooperation that we have described above are not only extremely common; they are often the very essence of how a new-economy company creates its market position. A company develops its market by investing in its ecosystem – sometimes even more than it invests in its own affairs.

Building partnerships in the ecosystem

The Fantastic Corporation was founded in Zug, Switzerland in 1996. The company was based on the idea that it would be possible to merge broadcasting with the Internet by developing software that could bridge all the gaps. This would remove many of the frustrations of both media.

Software development was well on its way by the end of 1996, and patents with 24 claims were filed in 1997. However, the key concern of the management was to ensure that the software would be received well when the first modules were launched over the following two years. This was critical, as the software represented radically new approaches to telecommunication.

One of the company's approaches was to invite leading companies from the industry to invest in it. This was done through a series of small capital increases, and Intel, Reuters, Lucent, Deutsche Telekom, BT, Telecom Italia, Loral Space & Communication, Singapore Press Holdings and Kirchgroup all took stakes in the venture. The strategy worked well, as Fantastic managed to build on the initial financial relationships to develop even more important commercial alliances with each of the nine partner companies. It formed joint ventures with Deutsche Telekom and BT to sell its software and services in Germany and the UK, respectively. Reuters agreed to provide content to its first trials and invested in Fantastic's subsidiary company in Japan. Kirchgroup, through its subsidiary Prisma Sports & Marketing, established a joint venture with Fantastic to deliver live and near-live video clips to next-generation mobile phones and cars. Singapore Press Holdings established a joint venture with Fantastic to deliver broadband media in Asia/Pacific, a project that paved the way for many new customers in the region. Loral offered corporate satellite services with Fantastic's software, Intel gave Fantastic considerable free marketing exposure at a time where it was crucial for the company, and Lucent licensed Fantastic's software and became a value-added reseller.

Goldmann Sachs took Fantastic public on the Frankfurt stock exchange less than three years after its incorporation.

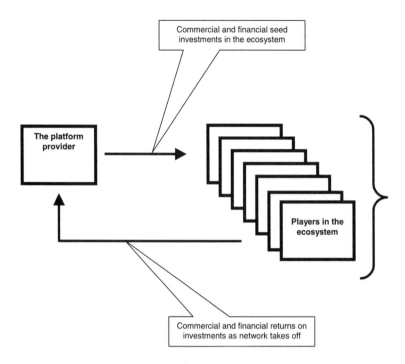

Figure 5.10 Investing in the ecosystem to build networks effects. A company operating in the digital economy needs to work closely with its commercial partners to generate the impact that eventually ensures it a network effect. The partner alliances may include joint development of standards, licensing agreements, financial cooperation or marketing alliances. The network effect sets in once the partners and customers start transacting with each other over the supplier's platform, and this creates an escalation of the supplier's business and revenues.

Responding to weakness in the ecosystem

We have studied a number of expansionary strategies, but what if our position in the ecosystem (for a product or the whole company) is so weak (or the market so dull) that it calls for harvesting or termination rather than for expansion? The typical solutions are here either progressive or defensive approaches. Progressive approaches are:

- *Comarket.* Combine your products with other weak products, possibly from other companies, to create a 'kennel of dogs' that can generate an attractive network or bundle effect.

- *Value added.* Combine the product with new features or other products to make it unique and/or more attractive. Many software companies have been particularly adept at relaunching software programs that initially failed commercially.
- *Reposition.* Position image deliberately as a product of the past to attract sentimental aficionados, or reuse it in a new context.

The alternative is the *defensive* approach:

- *Rationalize.* Cut costs and streamline operations. Although this may work, it may be only a temporary solution before losses reappear.
- *Harvest.* Increase minimum orders and cut all costs, including promotion, sales, service and research and development; to a minimum in order to milk before termination.
- *Graceful phase-out.* Merge the activity with another, then phase out the weak product.
- *Liquidate.* License the technology and/or brand, then stop producing/selling it yourself.
- *Sell.* Sell the company or unit.
- *Abandon.* Give the operation away or declare bankruptcy.

Whichever approach is taken, remember that the new economy is characterized by constant change and limited visibility, which means that any successful company will need to phase out or reposition products much more frequently than in the old economy. The way a company phases out can be almost as important as the way it launches. Joseph Schumpeter called it 'creative destruction'.

Audit Step 5: Competitor Audit

The review of the ecosystem in audit step 4 was based primarily on an overall evaluation of the company versus all other companies in the sector, and with a view to the possibility of forming constructive alliances. But some of these other companies will always be particularly important, and this calls for a competitor audit. A competitor is a company that aims to provide the same benefits to the same customers as your own company. It is also a company that you have decided to *treat* as a competitor.

The first step of the competitor audit is to define which market segments the company is competing in. Next step is to list the competitors by name. The third step is to segment the competitors into relevant groups. Fourth is the analysis of the most important competitors. Typical questions to ask should be:

- How strong is their management?
- How strong is their network?
- Does their business have barriers against competition?
- What are their overall strategies?
- How large are the resources at their disposal for the market?
- Have they implemented superior leveraging strategies?
- Do they have innovative advantages?
- Do they enjoy significant network effects?
- Do they have a lead in the development of standards?
- Do their offerings secure them switching costs?
- How strong are their development resources?
- Do they have critical patents?
- Do they have access to critical qualitative resources?
- Do they have access to sufficient capital?
- Do they fully understand the market?
- What is their business culture like?
- Do they have advantages/disadvantages in their cost structure?
- Do they have product advantages/disadvantages?
- Do they have service advantages/disadvantages?
- Are they mainly market oriented or product oriented?
- How large is their market share in different segments, and how is it evolving?
- Do they have a first-mover advantage?
- What is their image like?
- How do they promote and sell?

By following such an exercise, it will be possible to move to the next questions:

- What are their overall strengths, weaknesses, possibilities and threats?

- What are they likely to do next?
- Do they have strategic advantages that we can emulate?
- Are they posing threats that we can prevent?

After this analysis, we then work out what to do about the competitors. The first step here is to analyse how you should position your company towards them.

Positioning towards a specific competitor

Competitor positioning may take various routes, as the company can compete on product offering, development of network effects, leveraging tactics, speed to market, image, sales effort or promotion effort. Furthermore, it can choose to differentiate itself or to compete on all fronts.

What you choose depends largely on your competitive strength position. If the competitor has better products and better marketing, then you can't compete (and you will be positioned badly to cooperate because you will bring nothing to the table). So, you should either differentiate in order to get out of the uneasy position or you should do your homework by improving your basic offering. The opposite situation is where you have stronger products and stronger marketing, where it makes sense to compete head-to-head (Table 5.8).

There are several typical methods (some of which may be anti-trust violation if performed by a dominant player): if your company is in the situation such that it is so strong that you choose to compete against a specific company on all fronts.

- *Enforce incompatible standards.* This means seeking to enforce standards that do not fit the competitor well.
- *Block distribution chain.* This can be done by making exclusive deals or package deals with distributors, offering quantity rebates to distributors, or offering the same product in own brands and the distributors' brands ('private label').
- *Increasing distribution partners' switching cost levels.* This can be achieved by entering into production cooperation agreements

Table 5.8 Criteria for selecting competitive strategies

	Competitor has better offering	Competitor has weaker offering
Competitor has better marketing	• Improve offering • Differentiate • Develop a stronger network • Implement stronger leveraging strategies • Improve speed to market	• Improve image • Intensify sales effort • Intensify promotion effort • Cooperate • Compete head to head • Develop a stronger network • Implement stronger leveraging strategies
Competitor has weaker marketing	• Improve offering • Cooperate • Compete head to head • Implement stronger leveraging strategies • Improve speed to market	• Compete head to head

with the distribution partners, offering a very high level of service to distribution partners, establishing electronic ordering system or establishing ownership of storage facilities at distribution partners' locations.

- *Subsidize price on basic product.* For instance, offer software for free but charge for the support.

- *Subsidize price on part of the product.* For instance, give away software readers for free in order to create a pull, then charge for the corresponding servers.

- *Increase advertising.* Aim for economies of scale in the production through increased advertising.

- *Secure strategic technological advantages.* Maintain a high level of research or take out patents.

- *Prevent switching to other solutions.* Announce your plans for future improvements in order to prevent shifting.

- *Make exclusive supplier agreements.* Aim in particular at obtaining exclusivity on critical parts.

- *Signal the will to fight for the market share.* Announce aggressive sales targets and investments plans, or show willingness to follow competition in potential price wars.
- *Attract key staff.* Persuade key staff from the competitor to move to your side.

Audit Step 6: End-user Audit

While beating competitors can give some satisfaction, it isn't of course the primary marketing objective for a business. The primary objective is to create benefits for customers. Theodore Levitt made a great remark about how to view customers:

> The difference between marketing and selling is more than semantic. Selling focuses on the needs of the seller, marketing on the needs of the buyer. Selling is preoccupied with the seller's need to convert his product into cash; marketing with the idea of satisfying the needs of the customer by means of the product and the whole cluster of things associated with creating, delivering, and finally consuming it.

The purpose of the end-user audit is thus to develop the relevant understanding of customers, primarily through three steps:

1. Understand which real benefits your product provides the end users.
2. Divide the end users into segments that should be approached differently.
3. Determine how end users may be serviced efficiently on a one-to-one basis.

So, the starting point is to understand which benefits the company really provides for its customers. If you don't understand the benefits your product provides, then you can't promote it efficiently. Sometimes it can be useful to divide product benefits into the following categories:

- *Standard benefits.* Product characteristics that provide objective benefits, for instance, a car's ability to take the family safely from one place to another.

- *Corporate benefits.* Advantages other than the product itself that the supplier (you) takes to the customer. It may be the service or warranty agreements, or the fact that you can be expected to stay in business.
- *Differentiation benefits.* Advantages that make the product measurably better than specific competing products, for example, lower prices, a higher performance, longer lifetime or a better design.
- *Network benefits.* Advantages that the user will enjoy because of network effects associated with your offering.

Once the benefits are understood, it is time to move on towards segmenting the market. Any meaningful description of customers will be based on such a structured approach. This is true even in a world where each customer is treated individually. Segmentation creates the basis for understanding the business and targeting of messages, media and core market offerings. People or companies belonging to a significant statistical cluster comprise a segment of similar attitudes, habits and requirements; the better you understand which clusters your products sell to, the more targeted approach you can plan.

Segmentation may be based on criteria such as geography, demographics, social criteria, personal criteria, price sensitivity, purchasing situation, product usage or the benefits that the product provides (for further details, see Appendix B). It is important to understand that the benefit of segmentation is more than one-dimensional. It is more a conclusion that looks at the reasons why people buy, rather than a description of the purchase situation.

Once the segments have been defined, the company can choose the following marketing approaches (listed here in order of focus):

- *Aggregation strategy.* The least focused approach: the idea is to ignore the segments and approach the entire market in a uniform way. This is rarely a good approach, but if a company has a very strong short-term advantage that has to be utilized very quickly, then such a fairly simplistic choice may be preferred for some of its marketing activities.

- *Differentiation strategy*. Much more focused: the company chooses many or all segments, but also tries to approach them in different ways.
- *Niche strategy*. The company chooses one or a few segments and targets these intensively and individually.
- *One-to-one strategy*. The company treats each customer individually. Each customer is thus essentially viewed as a segment.

Few companies will pursue only one of these strategies. The marketing mix is usually a combination, with different elements of the marketing effort leaning against different segmentation strategies. However, the development of digital networking creates the possibility to move further and further towards fine segmentation, with one–to–one marketing as the most radical solution.

One-to-one marketing

The modern one-to-one marketing approach is based on digital networking, where computers continuously collect data about virtual visitors and customers, and then use this data to tailor the company's offerings. The concept was first described by Stan Davis and Bill Davidson in their book *2020 Vision: Transform Your Business Today to Succeed in Tomorrow's Economy* (1991). More comprehensive descriptions have since appeared in other publications, with Don Peppers and Martha Rogers as the leading pioneers. Their first book on the subject, *The One to One Future: Building Customer Relationships One Customer at a Time* was published in 1993, and they have followed up with numerous other books and articles.

Peppers and Rogers have highlighted a number of cornerstones on digital one-to-one strategies:

- *Identify each customer and visitor*. Ensure that the website (or other online environments) has the ability to identify all visitors and customers by building detailed profiles, which are improved each time the customer is online with the supplier. This works best if the customer is given a motive to provide information about motives, preferences, etc., so that the service can be optimized.

- *Customize for each customer.* Make sure that every customer receives a service that is unique. What is proposed to the customers should, for instance, reflect what they have chosen to look at and bought before, how they rate previous purchases, and any other information that they have volunteered about themselves.
- *Interact with customers.* A website shouldn't just be an electronic catalogue. The company should interact with the customers when they go online.
- *Protect privacy – and tell about it.* Inform customers that information that they provide will not be used for any purpose other than for customizing their services and that it will not be sold to other companies.
- *Describe how it works.* Explain to each customer how the interaction with the company works – how their data is being used to tailor the services for them, and what the benefits of this are.
- *Organize around customers, not products.* This approach makes it much easier for each customer to find exactly what he or she needs.
- *Enable each customer to gain some control over their profile.* Enable customers to correct any errors that a computer-generated profile might have created.

One-to-one marketing doesn't mean the end of other segmentation strategies. Quite the opposite in fact. A company that uses computers to handle relationships will generate superior information about clusters of users and will be able to use that information for superior segmentation that can be implemented in its marketing strategies.

Audit Step 7: Product Positioning Audit

The concept of the product positioning audit is one of the oldest exercises in business – building on the commodity school of marketing. But it has evolved considerably as the new economy has changed even the most basic ideas of what products really are and how product portfolios should be managed.

The new product paradigm

Let us look at the basic conception of a product. Throughout this book, we have used the term 'products' to describe what the company sells, since this is the most widely used expression in the industry. But we have to throw in a bit of confusion here. New-economy companies don't necessarily sell products. Since things change so fast and products are increasingly digitally connected, a product becomes a 'service' waiting to happen, and a product and a service combined is thus providing a 'solution'. Product marketing and service marketing approaches were described well by the original schools of marketing, but solutions marketing in a digital economy is new. Table 5.9 highlights some of the key differences in the philosophies. It shows how a product in the new economy is thus more and more likely to be connected and constantly monitored and supported by the supplier. It is designed, at least partly, by the customer, and it is engaged in networking both between customer and supplier and within communities of customers. Pricing is more fluent, and the product is changed continuously for as long as the customer has the need.

Product policies and path dependency

We have already described path dependency as one of the key characteristics of the new economy. But how does that relate to the product policy? This is largely a question of building in switching costs for the company's own products whilst reducing them for customers that use competing products. The typical switching costs are:

- the cost of learning to use new products;
- the cost of upgrading to new versions;
- the cost of entering data and personal set-ups in new products;
- the cost of finding new suppliers;
- the cost of paying for alternative products;
- the loss of loyalty benefits in a switching situation.

Many of these parameters apply for old-economy products as well, but one in particular is important for digitally digital products: the issue of

Table 5.9 Comparison between product approach, service approach and solution approach

	Product approach	Service approach	Solution approach
Key competitive factors	Price, quality, delivery, convenience	Ongoing support	Ability to upgrade and maintain online relationship with customer
The supplier's time horizon	Time of the product sale ('the moment of truth')	Duration of the service contract	Duration of the customer's need
Dominant cost focus	Direct manufacturing and distribution costs	Period costs	Design and marketing costs
Product support approach	One-year warranty	Call the supplier if there is a problem	Supplier informs customer if they see a problem
Design approach	Fixed and uniform within product portfolio	Customized for each user	Customized by each user
Revenue model	List prices	Subscription period prices	Individual and fluent prices
Key to customer's sales experience	Presentation	Relationship	Involvement
Key marketing objective	Brand loyalty	Relationship building	Community building
Key competitive resources	Capital and production facilities	Staff	Customer networks obtained through first-mover advantage
Essential source of value	Manufacturing and distribution process	Training and maintenance process	Leading platform status and network of interacting users

It is still convenient to use the term 'product' for what the company delivers to the customer, but the approach to products changes as they are combined with services and connected through digital networks. This can be called a 'solution approach'.

versioning. The key here is how the company manages compatibility – the ability for one version of its product to work with other versions, and with competing products from other suppliers. Table 5.10 shows the different approaches to compatibility.

Table 5.10   Four strategies for compatibility at product migration

	Control	Openness
Compatibility	Controlled migration	Open migration
Performance	Performance play	Discontinuity

Controlled compatibility means that the customer can migrate from one product to another, but only in directions that the suppliers control and within proprietary technologies. Typical approaches are to enable users of older versions of your own product and users of competing products to migrate to your own, but to try to prevent the opposite. One-way compatibility is commonly used. Microsoft's Office 97 could, for instance, read files from Office 95, but the reverse was not possible. *Open migration* is when the company enables any migration path within an open technology. This approach, which is the most user-friendly, makes particular sense if the supplier's key advantage is manufacturing and marketing skills. *Performance play* is the launch of a new product that is neither backwards compatible nor compatible with competing products. This approach makes most sense for a company with a very strong position. *Discontinuity* is a move towards a radically different concept that leaves previous technologies outdated. This strategy often requires a very big improvement in technology. How big a performance advantage must you offer if you want to launch without backwards compatibility? Both Peter Drucker and Andy Grove have spoken about the '10 times rule' – new technologies must be at least 10 times better than what they replace in order to have a chance of success.

Product portfolio approaches

The new economy is less predictable than the old one, and this means that it is more difficult to predict the marketing success of any new product. This is reminiscent of the situation in another volatile industry, financial markets, and the conclusion is the same. When a company operates in volatile industries, it can't bet on any single solution. It has to take a portfolio strategy. Software people have many expressions for this. They talk, for instance, about 'throwing things up on the wall to see what sticks', or the 'seed, select and amplify' strategy. You start a number of things, select the ones that seem to work and amplify those. This doesn't sound very scientific, but it does reflect a realistic attitude to markets dominated by feedback loops and path dependency. It also reflects a realistic attitude to failure. Many of the most successful companies in the new economy accept that employees may be great performers even if they are involved in a string of commercial fiascoes. A company is judged not by its results, but by what it did in the process. The same goes for the judgement of managers and entrepreneurs.

The 'seed, select, amplify' approach is fine, but it doesn't mean that the company shouldn't have a strategic approach to products. The first step in development of such an approach is to evaluate each of the products in terms of their market position, for example, by using the Boston Consulting Group Matrix (Figure 5.11).

The products are plotted into this matrix based on their market share within the relevant market and the overall growth of that market (here, 'market' must be understood as the market area that the product is realistically competing in). The 'Question mark' represents the situation in which the company needs to gain market share in order to benefit from network effects, increasing returns and first-mover advantages (path dependency).

'Star' is a position where the company is likely to make money and where it makes sense to invest. Because of the high market share, the company is likely to enjoy network effects and increasing returns, which again means that it is able to substantially fund market and product development and that it will be able to tolerate a price war better than some competitors. The market growth means that it makes sense to invest in the future. There is a potential trap in this position, however. If the market is growing rapidly, then the company may experience rapidly rising turnover but a falling market share at the same time. Falling market share is always a danger signal.

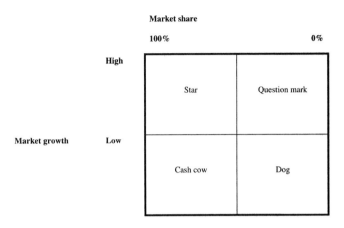

Figure 5.11   The Boston Consulting Group Matrix. This matrix is one of the most widely used graphical models in marketing. It describes how different products in the corporate portfolio are positioned in terms of market share and market growth. It goes without saying that market growth is a key criteria in evaluation of a product's position. Market share has gained importance as digital connectivity creates powerful network effects for market leaders. The terms 'star', 'question mark', 'cash cow' and 'dog' were created by Boston Consulting Group.

'Cash cow' is a situation in which the company will typically make money, but where it does not make sense to invest very much. The best option is often to prepare to phase out the product while building on the network effects for the launch of new products. 'Dog' is typically an unsustainable position, where the best decision could be to sell, reposition or even terminate the operation.

Most products are launched into markets with a high growth rate, and as they mostly start with a low market share, they must be classified as 'question marks'. From this dubious position, they will typically either move to 'star' or 'cash cow' and then 'dog', or they will move straight to 'dog'. From 'dog', they will often be eliminated or redefined so much that they actually target a new market and can re-enter the cycle.

We shall come back to the Boston Consulting Matrix when we address Step 11 of the marketing audit. This is all about speed and timing in the digital economy, which is very relevant for how the company's product portfolio evolves over time.

Audit Step 8: Pricing Strategy Audit

One of the marketing areas that has experienced the greatest impact from the development of digital networking is pricing. How, for instance, should a company price something that costs a fortune to develop and market, but costs nothing to replicate? And how should it price the product if more and more users know everything about its competitors' prices and if borders don't mean anything any more? Let's look at 25 typical pricing strategies in the old and new economies. The first 12 are based on product-based pricing criteria:

1. *Quantity rebates.* Offer linear or progressive discounting as customers buy higher quantities.
2. *Cost-based pricing.* Price the products at cost plus a mark-up.
3. *Value-based pricing.* Price on the basis of the perceived value that the product has for its user.
4. *Price signalling.* Sell at a high price in the expectation that customers will conclude that the product has a high value due to the high price.
5. *Price bundling.* Offer discounts for purchase of several products.
6. *Premium pricing.* Subsidize low-end/entry products and earn more on high-end/premium products.
7. *Complementary pricing.* Subsidize the core product and earn more on the supplies.
8. *Supply-controlled auction pricing.* An auction for a product that is supplied from a single source.
9. *Free pricing.* Give away a product in order to build a network effect and a platform status.
10. *Negative pricing.* Pay people to receive a product or service, for instance for clicking on ads. This assumes that other customers are paying for the result.
11. *Flat rate pricing.* Charge a single fee for unlimited, variable use of an asset that has close to zero marginal costs.
12. *Conversion pricing.* Provide the product at a cheaper cost if the customer upgrades from an earlier or competing version.

The next eight strategies are based on market-based discrimination criteria:

13. *Geographical pricing.* Sell at different prices in different geographical areas.

14. *Second market discounting.* Sell the same product in two different markets or under primary and secondary/no brand. Second market is priced lower.

15. *Membership pricing.* Offer discounts to club members.

16. *List pricing.* Enforce a fixed price that retailers may not deviate from.

17. *Sole-supplier discounting.* Give a discount if the customer will agree to make you their sole supplier of the product category.

18. *Differential pricing.* Charge different prices depending on who the customer is.

19. *Demand-controlled auction pricing.* Auctions by consortia of buyers who look jointly for the best bid for large, combined orders.

20. *Competitive pricing.* Price at any time is just below what your key competitor is charging.

The last five strategies are based on the timing factor:

21. *Penetration pricing.* Launch at low price in order to achieve high market share and move towards network effect and increasing returns.

22. *Skimming pricing.* Launch at a high price, then reduce it.

23. *Periodical discounting.* Discount the product from time to time, typically during seasonal sales.

24. *Random discounting.* Discount to get volume, but at unpredictable times so that potential buyers cannot plan to buy at the next discount period.

25. *Delayed version pricing.* Charge less for a delayed version of an electronic product.

The move from old economy towards new economy has changed the relative dominance of these different pricing strategies. The following strategies have become more difficult to support as digital connectivity has emerged:

- *Skimming pricing.* This is still common, even in the high tech industry, but it can be detrimental to an attempt to build a network effect. The most common version of the strategy will be one in which the price is dropped gradually, not as an attempt to reach more customers, but as a response to the fact that new and better versions of the product have been launched. Intel chips are a good example.

- *Cost-based pricing.* This does still exist, but it makes little sense if the marginal costs are extremely close to zero.
- *Geographical pricing.* It will become more difficult to differentiate on the basis on geography as we gradually have full pricing transparency due to electronic networks.
- *Differential pricing.* It will also become more difficult to differentiate prices based on who the customers are if the customers obtain full transparency of pricing via electronic networks.

Other pricing strategies are gaining importance, however:

- *Penetration pricing.* This makes tremendous sense in any scenario with increasing returns.
- *Delayed version pricing.* Digital products can often be launched at different times. An example is delayed stock prices.
- *Quantity rebates.* Again, a strategy for network effects that is easy to support if marginal costs are low.
- *Value-based pricing.* This is as opposed to cost-based prices. Makes sense if the value is very different from the marginal cost price.
- *Price bundling.* This is often used when a buyer has separate budgets for capital and operating expenses. If capital investment budgets are tight, then move some of the pricing to service. Another example is software that is free, but the customer pays for the manual.
- *Premium pricing.* Strong strategy for creating a platform that can be the basis for subsequent launches of premium products.
- *Complementary pricing.* Same as for premium pricing.
- *Supply-controlled auction pricing.* This is the typical Internet exchange model. It doesn't make sense to make auctions over information products with very low marginal costs, but it does make sense to use digital communication infrastructure to make auctions for limited-supply products.
- *Free pricing.* An aggressive strategy for network effects.

- *Conversion pricing.* Frequently used for software products.
- *Negative pricing.* This (fairly rare) strategy is typically used as a means to get people to click on ads.
- *Flat rate pricing.* Makes sense for online connectivity where the marginal costs are minimal.
- *Demand-controlled auction pricing.* This is the situation where several buyers 'meet' each other over a digital network and join forces in the search for a good joint offer.

A company's pricing policy will always be questioned by its customers, and it is thus vital that it is in line with the overall corporate strategy and that the company can argue well for it.

A case of complementary pricing

On 1 October 1998, SkyDigital was launched to become the first digital TV platform available in the UK. Digitalization of the distribution offers the distributor and consumer several benefits. However, the costs were high, as the project required major investments in set-top boxes to convert the digital signal into an analogue signal that current TV sets could understand. These boxes cost between £150 and £250.

Sky charged for the boxes in the beginning, but then they changed their strategy and began offering them free. Why? With digital distribution and set-top boxes, they could offer other services, such as shopping, home banking, email, information services and games. With access to a large customer base, Sky would be able to make a lot of money from these services, so they decided not to ask for a monthly rental fee for the set-top box.

What were the results? Shopping services on the SkyDigital platform were offered under the name 'Open'. Just nine months after launch, Open enjoyed considerable success:

- It became the largest e-commerce platform in the UK, with access to over nine million people through TV sets in over 3.4 million homes.

- Research showed that during June 2000, 1.6 million homes used Open and more than 1.1 million homes used it at least once per week.
- Over 267 000 Open homes (11% of average installed homes) had made a purchase through Open since October 1999; of these, 35% had made repeat purchases.
- BSkyB anticipated that by 2005, the Open service would generate gross interactive transactions of approximately £350 per subscriber per annum, contributing to total revenues of approximately £50 per subscriber per annum.

Lessons learned:

- Customers are reluctant to pay for hardware. By giving out the set-top box for free, Sky had over 1½ years converted 40% of their existing customers into digital users. By choosing to either rent or sell set-top boxes to their customers, operators in other countries succeeded in converting only 5% of their customer base.
- It is not worthwhile for operators to charge for hardware. Instead, look for the new revenue streams, and make sure that you tap into this before your competitors do.

Audit Step 9: Distribution Strategy Audit

The distribution strategy includes international market selection process and chain strategy.

International market selection process

Some of the typical reasons why a company may choose to expand abroad are:

- *Network effect and increasing returns.* Expanding aggressively ensures the necessary leadership.

- *Desire to diversify risk.* Presence in many markets reduces risks (political, economic, currency exchange rates, etc.) associated with any single market.
- *Access to technology.* International presence may ensure access to new technologies and approaches evolving abroad.
- *Movement of customers overseas.* Obtain ability to service the same customers internationally.
- *Access to resources.* Produce where the cost is lower at any given time.
- *Culture.* Develop an international corporate culture.
- *Acquisition.* Possibility to acquire competitor abroad.
- *Spirit.* Maintain an entrepreneurial spirit after local dominance has been achieved.
- *Distribution structure.* Deliver to distributors with international network.

Two terms are important for the discussion of the way companies select their international distribution: 'passive market selection', which means that the company passively picks up the distribution opportunities that it comes across, and 'active market selection', which means that the company identifies and approaches what it believes are the best markets. Active market selection is again divided into 'expansive' and 'contractive' selection. Expansive selection is a strategy whereby the company expands into new areas that are close to, and resemble, the areas it already knows. Contracting selection is a strategy in which the company maps all potential market areas and then eliminates those that appear least attractive. The result is a short-list of the most attractive markets. The company will typically change its selection process as it grows (Figure 5.12).

The strategic selection of focus markets has not changed radically in the new economy, but chain strategy has.

Chain strategies

The typical alternative international distribution strategies are either direct or indirect. Direct distribution may be achieved through direct

Revenues

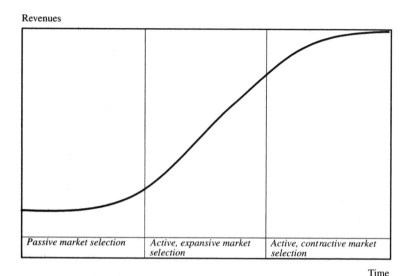

| Passive market selection | Active, expansive market selection | Active, contractive market selection |

Time

Figure 5.12 Company growth and international market selection process.

sales from a website or a similar direct electronic communication system. Indirect sales can be through portals or digital exchanges.

The electronic distribution opportunity deserves a few comments here. A company can distribute electronically by setting up a website or another electronic infrastructure (e.g. for cars, mobile phones, digital TV, kiosks). When doing so, it will often be able to handle the distribution tasks far more efficiently than it could through physical distribution.

Electronic distribution can, like physical distribution, be either direct or indirect (Table 5.11). Direct electronic distribution involves setting up a website (or similar for mobile phones, cars, set-top boxes, etc.) and enabling customers to purchase in this way. Indirect electronic distribution involves engaging in commercial agreements with other electronic sites (digital exchanges or portals) that provide hyperlinks to your site. Hyperlinks in essence fulfil many of the same functions that the wholesalers and retailers did before: they constitute a distribution network, which means that the company's own website, in some cases, work more as a back office than a distribution system.

Table 5.11   Comparing physical and electronic distribution

Task	Physical distribution	Electronic distribution (corporate website, portal or digital exchange)
Providing product comparisons and ratings	Producer has a much better understanding of the offer than the buyer does. Information passed to the buyer depends on dedication and time of salesperson and is biased by the availability of stocks in the shop	Buyer has access to transparent product comparisons and ratings by external and neutral sources
Providing and adjusting prices to market conditions	Prices change rarely since it is complex to communicate the change and handle repricing of stock	Prices can be changed frequently or even constantly
Providing search and filtering facilities	Search and filtering is limited to what is available in print catalogues and the advice a retail salesperson might provide	Search and filtering can be virtually unlimited
Providing new product alerts	Alerts take place through in-shop promotions and direct mailings	Alerts take place via email and on home page. They can be individualized so that buyers only sees things that they are likely to be interested in
Providing access to product-related discussion forums	Electronic bulletin boards are not part of physical distribution solutions	Full access to product-related discussion groups where superior and unbiased advice can be obtained
Provide recommendations based on users' search and purchase habits	Recommendations depend on personal relationship with a specific salesperson, who may or may not recall the users' previous purchase patterns and stated preferences	Software programs perform statistical correlation analysis to predict what the users might be interested in
Provide electronic payment services	Payment can be made with cash or credit card at the counter	Payment is made electronically
Provide shipping services	Users can carry products with them or, in some cases, have them shipped	Products are shipped or downloaded electronically
Provide electronic product presentations	Products can be inspected physically (but only the samples that are in the shop at the given time)	All products can be inspected electronically but not physically
Provide product-related information, such as news, regulatory data, research, reviews, awards, etc.	Generally very limited or not available	Unlimited possibilities
Provide information about the buyers' previous purchases, wish lists, shopping baskets, etc.	Generally not available	Generally full transparency
Provide digital exchanges linking any buyer to any seller within product categories	Generally not available	Can be unlimited
Provide data for advertisers	Data can, primarily, be collected on sampling basis	Data can be collected in real time and based on the entire universe of clients and surfers
Coordinate flow from suppliers via manufacturers to distribution	An order from an end user triggers orders from manufacturers to suppliers	An order from an end user *is* an order to the ultimate supplier

A common misconception at the early days of the Internet boom was that it would affect mainly distribution by eliminating the middle people. The reality is different. There is still a great need for these intermediaries, but they are serving new functions that couldn't be provided before.

Audit Step 10: Promotion Strategy Audit

Promotion is a term that covers advertising, public relations and sales campaigns (Table 5.12). The promotion strategy of the company should first and foremost cover:

- identification of the communication target groups and messages;
- branding strategy.

Table 5.12 Overview of promotional parameters

Advertising	Public relations	Sales campaigns
• Own corporate websites	• Press releases	• Discount offers
• Cross-branding agreements with other websites	• Press kits	• Promotion activities in shops
	• Newsletters	
• Portals	• Participation in committees	• Coupon arrangements
• Digital exchanges		• Copackaging
• Printed media	• Give interviews	
• Television	• Make speeches and presentations	
• Radio	• Press conferences	
• Brochures	• Participation in fairs and exhibitions	
• Demos		
• Email campaigns	• Stimulate third parties to write analysis of your offerings	

Determining promotional target groups and messages

Promotional activity must be targeted at its audience. This may include corporate stakeholders, such as shareholders, strategic partners, and potential and existing staff. It should also address distribution (physical distributors, website, portals, exchanges), influencers (advisors, suppliers of complementary products, industry gurus, aggregators), end users and the press (analysts, trade press and mass media).

A company will often start out with a rather simple promotion strategy, but as it evolves, it will invariably segment the target groups and refine the message more and more. The key questions to ask here include:

- Who are our key promotional target groups?
- Why are they interested in your company/products?
- When do they need information from us?
- What is our desired corporate image?
- Why do our distributors (if any) sell our product?
- When do our distributors make decisions about distributing ours and competing products?
- What benefit does our product give the distributors?
- How is our product positioned as compared to competing products?
- What is our product image like, compared to those of our competitors?
- How can our distributors and end users be segmented or addressed on a one-to-one basis?
- How do our competitors communicate?
- How mature is our product in the product cycle?
- Which third parties influence our customers in their purchase decisions?
- How, when and where do they exert their influence?
- Where do our end users buy the product?
- Why do our end users buy the product?
- When do our end users buy the product?
- When do our end users use the products?
- What benefit does our product give the end users?
- What message do we want the press to spread?
- Which target groups do we want the press to spread the message to?
- What is the focus of the message?
- Where and when shall the press be approached?

Identification of target groups and messages is not very different in the new economy to the old one. You segment the target groups and ask a number of essential questions that lead to the development of a set of strong and comprehensible messages. It is the craft and art of marketing promotion that has been described in rich detail, specially by authors of the commodity school and the managerial school of marketing. The branding game, on the other hand, has changed radically in the digital economy.

Branding strategy

To understand how branding can be done, try to think of two different product experiences. A car contains parts from thousands of direct suppliers. But how many brands does the user see? The user sees the name of the car manufacturer (aggregator, if you will), maybe the name of the tyres, and sometimes the name of the provider of the radio. The aggregator has reduced everyone else to a commodity provider. Next, go to the Internet and surf around for an hour. How many brands do you see? Chances are that you see hundreds.

The branding game is different in the new economy because of some of the basic characteristics of technology and the Internet. Digital technology is, like a car, based on combinations of numerous components. In the software and telecommunication business, these are often described as 'stacks' of software. Some of the providers of these layers are anonymous, but many others are not, because what they provide is not a commodity. It is a de facto standard, protected by a network effect and perhaps patents. They can insist on branding rights.

The second difference is that the core concept of the Internet is the ability to hyperlink. You are one click away from virtually anything, so while it would be nice for any player to suppress other brands, it is in reality almost impossible. So, the companies do something different. They exchange brands on commercial terms, offering to create a link to a supplier on their website, if they will do the same. The key asset in this game is the digital real estate (i.e. screen images) that they control. This means that the key to the branding strategy is to maximize the digital communication that flows over the platform (eyes) and to leverage this by exchanging branding with partners.

Audit Step 11: Speed and Timing Audit

Speed and timing are often described as strategic marketing parameters in the traditional marketing theory, but they are rarely claimed to be very important. In the new economy, not only are they important, they are often *essential* to the marketing exercise.

The reasons are simple. The new economy is, more than anything, about reaching leadership, and getting into the increasing returns territory before the competition does. So speed is critical. So is timing. New-economy companies operate in sophisticated ecosystems where they work with several other companies, and the sequence of events that they choose can mean everything.

Organizing for speed

One of the authors that has written most about the speed issue in the new economy is Peter Drucker. This is rather impressive, since he was 85 years old when the Internet took off. In 1997, Drucker was featured on the cover of *Forbes* magazine under the headline, 'Still the Youngest Mind', *Business Week* has called him 'the most enduring management thinker of our time'.

The title of Drucker's book, *Post-Capitalist Society*, refers to his observation that the digital economy really has changed the key resource in society from capital to know-how. It is not so much the money that rules, Drucker said, as the brains. The book described many of the steps a company can take to be able to manoeuvre faster in the new economy. Drucker pointed out that it is key that you mobilize every individual to think: 'The modern organization cannot be an organization of "boss" and "subordinate"; it must be organized as a team of "associates" . . . The task of management in the knowledge based organization is not to make everybody a boss, it is to make everybody a contributor.'

Drucker also insisted that the fast-moving organization also has to learn to rethink very quickly: 'It has to learn to ask every few years of every process, every product, every procedure, every policy: If we did not do this already, would we go into it now, knowing what we now know? . . . Every organization requires continuing improvement of everything it does.'

A part of the speed and timing audit is thus to go through all functions of the company with the sole focus of identifying every reason that might slow things down – and, if at all possible, eliminate each of these factors.

Mastering timing

Traditional marketing literature describes a number of typical phases in the product lifetime. A simplified version is shown in Figure 5.13.

Market phases

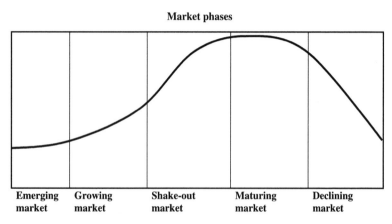

| Emerging market | Growing market | Shake-out market | Maturing market | Declining market |

Figure 5.13   Phases of the traditional market cycle.

Software for computer chess does not use the same software algorithms for the opening game as it does for the mid-game or the closing game. Nor can management of a company use the same approaches during each step of the five-market cycle. It might not even be the same management team that is able to excel in each of the stages (Table 5.13). During the introduction and growth stages of the market, you will see entrepreneurial management who can sell, motivate, inspire and force rapid change. Here, personality and charisma are the keys to success. This is also a high-pressure environment, where focus is on shipping the first products and raising capital through different

Table 5.13 Management focus over the market cycle

	Introduction	Growth	Maturity	Decline	Termination
Management profiles	Entrepreneurial	Entrepreneurial	Professional fine-tuning	Professional fine-tuning, administrative	Administrative
Management resource pressure	High	High	Medium	Low	High
Corporate pressure points	R&D Finance	Engineering Marketing Finance	Production Marketing	Finance Marketing	Finance

Management style, critical resources and pressure points change considerably over the market cycle.

forms of leverage. As the market matures, there will be an increasing need for 'professional' skills: experience in fine-tuning the marketing mix, handling complex negotiations, and structuring organizations. Focus moves towards engineering, marketing and production as the first products are launched. In the last phases, there will be a need for rationalization and cost-cutting, which appeals mostly to management with strong administrative and financial skills.

The way the company gathers information for its planning purposes will also change over the market cycle (Table 5.14). The initial phases will be based primarily on explorative marketing research while quantitative approaches, including digital tracking on websites, increase

Table 5.14 Marketing research methods over the market cycle

	Introduction	Growth	Maturity	Decline	Termination
Market analysis tools	Focus groups Informal interviews Key interviews Supply mapping Informal interviews Online market research	Focus groups Informal interviews Key interviews Supply mapping Informal interviews Online market research	Concept tests Product tests Concept/ product tests Advertisement pretests Packaging pretests Online market research	Consumer panels Distribution research Focus groups Informal interviews Market maps Radar Retail panels Supply mapping Online market research	None

Companies will use different elements of this toolbox in each stage of the market cycle.

in importance over time (see Appendix C for an overview over the most useful field research methods).

The products also change considerably over the cycle (Table 5.15). Products are initially launched as stand-alone solutions ('ugly boxes' if they have a physical element) with basic functionality. A number of new features and variants are introduced during the growth phase. Here, we may also see the product converging or becoming embedded in, and connected to, other products. The mature market will see rapid increases in quality and further convergence and embedding into existing products. The last phase may evolve in many different ways. The product might disappear due to lack of interest or (more likely) because a better solution is made available. Alternatively, the product may simply be swallowed completely by another product

Table 5.15 Product characteristics over the market cycle

	Introduction	Growth	Maturity	Decline	Termination
Product characteristics	Stand-alone Basic Low quality	Second generation Initial convergence Connected Medium quality	Segmented Converged Connected High quality	Embedded	Embedded Stripped

Frequently in the new economy, products become embedded and reduced to features as market cycles mature.

(sold by other companies), and sold so that it is now simply reduced to a feature.

The embedding phase in the mature market is often critical for the end game. Embedding is easier and much more common in the digital economy than before, and it is often the market leader that wins by embedding relating products through aggressive complementary pricing strategies.

Convergence and embedding of products is related closely to the development of alliances (Table 5.16). The early stages are often characterized by a focus on definition of product standards. There will also be other alliance activities, but these tend to be fairly loose and informal, as the market situation is too fluid to form long-lasting, binding agreements. Product pre-announcements and joint PR activities are also common at this stage. The subsequent alliances will be mainly for comarketing (which are often quite simple to implement). During the growth stages, there will also be an increasing number of production-related alliances, where companies exchange APIs and start to integrate their products more closely. During the final stages, there will be a large number of mergers and acquisitions as players rationalize their activities and embed and converge product offerings.

Table 5.16 Marketing alliances over the market cycle

	Introduction	Growth	Maturity	Decline	Termination
Alliances	Informal marketing alliances Participation in standards committees and alliances	Participation in standards committees and alliances Informal marketing and production alliances	Formal marketing and production alliances Mergers and acquisitions	Mergers and acquisitions	Mergers and acquisitions

Marketing alliances are critical to success in the new economy. The early stage of the market cycle is characterized by many loose alliances as they firm over time. The end game has many mergers and acquisitions.

One of the most critical aspects of the new economy market cycle is the targeting of customers (Table 5.17). The company will typically target small niche areas initially, and then increase its scope as it grows. However, the more critical aspect is perhaps the attitudes that characterize the customers at different stages, for example, innovators, early adopters, early majority, late majority and laggards.

Table 5.17 Target markets over the market cycle

	Introduction	Growth	Maturity	Decline	Termination
Number of segments	Very few	Some	Many	Many	Some
Customers	Innovators	Early adopters	Early majority	Late majority	Laggards

One of the most critical aspects of the market cycle in the new economy is the transition from selling to innovators and early adapters to selling to early majority.

The innovators in the new economy are *technology enthusiasts*. They are the people who like new, technical things because they are new and technical. The early adopters are the *visionaries* who can see the commercial and technical potential and want to enjoy it before their competition. The early majority are more *pragmatic*. They join because it makes sense, but not until they are pretty sure that it works. The late majority are *conservative*. They join the revolution only because it has become obvious that they have to. And then there are the laggards, the *sceptics* who were against it all the way but who finally have to give in.

No one should study marketing in the new economy without reading Geoffrey E. Moore's book on this phenomenon. In *Crossing the Chasm*, Moore describes one of the biggest timing challenges new companies in the new economy have: 'The transition from selling to enthusiasts and visionaries to selling to early majority is not continuous.'

We can call this 'Geoffrey Moore's Law' (Figure 5.14). The issue is that there is a huge chasm in the market cycle that any company can fall into. The chasm is there because the pragmatists will have completely different product requirements to the enthusiasts and visionaries. They expect rich applications, communities, service, support, etc. They expect a solution that is far more complete and more demanding to create and deliver, than what the early buyer would accept.

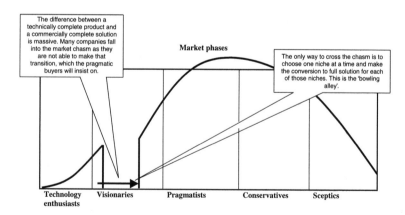

Figure 5.14 Geoffrey Moore's model for the discontinuous market cycle. Geoffrey Moore described how many companies struggle to make the huge transition from providing a product to providing a full solution that can satisfy the pragmatic buyers.

The way around the problem is to select a few application areas and market niches and then create the full solution for these target areas. Moore calls this the 'bowling alley', whereby the company crosses the chasm with one application at a time until it has finally crossed the chasm as a company and reached the mainstream. This brings it into the 'tornado' phase, in which the market feeds on itself in a process of virtually automatic escalation.

Distribution and promotion strategies are also very different in the different market phases (Table 5.18). Distribution will initially be based on a 'push' strategy: focus on stimulating distribution chains to carry the product. This can be physical distribution and electronic distribution though portals and exchanges. Later on, there will be increasing focus on 'pull' – stimulating the demand from the end users. In the late stages, a company may resort more to push again as demand diminishes.

Table 5.18   Distribution and promotion strategies during the market phases

	Introduction	Growth	Maturity	Decline	Termination
Distribution strategy	Channel incentives	Channel/ end-user incentives	Channel/ end-user incentives	Channel incentives	Little/no incentives
Promotion strategy	Create awareness	Provide information Create positive attitude	Create confidence Remind	Remind	Remind

Distribution and promotion strategies change from a push focus to a pull focus over time.

The promotion strategy changes in similar ways. The initial efforts will be centred around creating awareness, particularly so in the distribution chain. Product pre-announcements are often key. Later on, there will be a requirement to provide information and create a positive attitude for the activities. During the shakeout phase, there is a need to create confidence as customers can see that there is a shakeout going on. This is also the phase in which the company will increasingly remind people of the products. Reminding is the promotion focus during the decline and termination phases.

The final aspect to consider in the timing audit is the financial issues (Table 5.19). The company will initially experience huge losses, and needs to maximize its leverage while reducing risk. This is done by issuing stocks and stock options. If the business model is proven, then it can use more traditional leveraging methods, such as issuing bonds and taking bank loans. Revenues will be very low or nil in the early phases, as the company aims to establish itself as a leading standard and as it suffers from the lack of network effects. Cost will typically be a secondary consideration in the early phases, as the focus is on growth and development of network effects. However, as the market matures, it will become easier to find cost reductions through economies of scale and minimal marginal costs, and because the company is going through a learning curve. Cost reductions are key in the late phases (again, unless there is considerable milking and monopoly activity).

Table 5.19 Financial characteristics of each market cycle stage

	Introduction	Growth	Maturity	Decline	Termination
Funding	Issuing stocks to Business angles and venture capital Issuing stock options	Issuing stocks to venture capitalists and strategic investors IPO	Bank loans and corporate bonds Spin-off activities	Bank loans and corporate bonds Spin-off activities	
Revenues	Very low or nil	Increasing	Massive	Stalling	Declining
Gross margins	High	High	Medium	Low	Low or negative
Cost reductions	Few	Some	Many	Many	Few

The appropriate financial tools change character as the business segment moves from high risk/high potential towards lower risk/lower potential.

Defining the Direction for the Company from the Marketing Audit

Once the company has been through the 11 phases of the marketing audit, it is time to try to draw the conclusions. A number of steps have proven to be useful:

The ideal portfolio

What if the company has an unbalanced product portfolio according to the Boston Consulting Group Matrix? Table 5.20 illustrates the typical problems associated with such a situation.

Table 5.20 Problems associated with an unbalanced portfolio

	Only question marks	Only stars	Only cash cows	Only dogs
Cash flow	Too high requirements for capital	Good, but large requirements for investment capital	Good	Acceptable or negative
Profit	Negative	High	Good	Low or negative
Growth	Potentially good	High, but the company is likely to stall soon	None or negative	Negative
Demands on management	Too high	Too high	Low	Too high
Stability	Low	Satisfactory	High	Varies

- Make a SWOT (strengths, weaknesses, opportunities, threats) analysis.
- Define key competencies.
- Define lasting competitive advantages.
- Define key strategies.
- Determine strategic gap.

The SWOT analysis is a simple procedure for collating what you have concluded from the 11 steps of your marketing audit: simply list your main conclusions as strengths, weaknesses, opportunities and threats in a table (Figure 5.15).

It is often possible to change weaknesses and threats into possibilities. For instance, if your company is weak on a given product, it may choose to terminate production of that product and form a

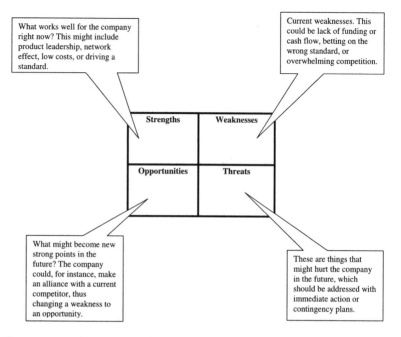

What works well for the company right now? This might include product leadership, network effect, low costs, or driving a standard.

Current weaknesses. This could be lack of funding or cash flow, betting on the wrong standard, or overwhelming competition.

Strengths | Weaknesses

Opportunities | Threats

What might become new strong points in the future? The company could, for instance, make an alliance with a current competitor, thus changing a weakness to an opportunity.

These are things that might hurt the company in the future, which should be addressed with immediate action or contingency plans.

Figure 5.15 The SWOT analysis for summarizing the essentials of the marketing audit.

marketing alliance with a company that is strong on this product but weak in areas that your company is good at. This product will thereby move from being your weakness or threat to being your strength or possibility.

In order to define the strategies, you can move quickly to the partial analysis again to check if there are elements in these conclusions that must be adjusted before you write the marketing plans. Once the SWOT conclusions have been listed, the company must seek to conclude what its key competencies are. This can be achieved by comparing the SWOT conclusions with information about the competitors. The question to ask is: 'What do we fundamentally do better than the others?'

Once the key competencies have been defined, it is interesting to consider whether these are under threat. In some companies, they will be connected closely with the capabilities of a few key people, and if these people leave, then the company will be under threat.

They may also relate to technologies that may become obsolete because of new innovations, or to key distribution patterns that may not be sustainable.

The third step is to define lasting competitive advantages. These are advantages that the company believes will enable it to continue to remain competitive. They are the pillars of the company's future. We then move on to describe key strategies to summarize:

- the markets you will penetrate;
- the value strategies you are pursuing – and why;
- the products you will offer;
- how you will produce and market them;
- how you will build your network;
- the kind of partners you will have;
- the kind of competitors you will have;
- how you will use leverage;
- where you expect your competitive barriers to come from.

Finally, it can be useful to determine the difference between what the company does now and what it needs to do in order to implement the strategies it has chosen. Perhaps the organization or core resources are insufficient? The gap analysis should highlight this, thus allowing the company to implement steps to close the gap.

WRITING STRATEGIC MARKETING PLANS

As we have now been through the thought process in the strategic marketing, we can move on to what is normally much easier: writing the plan. Appendix D gives an overview of the different steps of the strategic marketing audit, and how they can be relevant as input to different parts of the strategic marketing plan.

Long-term marketing plans will often have a structure as outlined in Table 5.21. However, no fixed template can ever be made; each company will have to develop its own.

Table 5.21 Structure of long-term marketing plans

Section	Examples of what may be included
1. Executive summary	This is typically one to two pages, which summarize the most important observations, threats, opportunities and conclusions. The summary should refer to the company's mission, business segment, value-creation strategies, key competencies, main marketing strategies and main marketing parameters
2. Analysis of the business environment	The purpose of this analysis is not to prove how much you know about the market, but to extract the essential underlying currents in the surroundings and their strategic consequences
2.1 The market	This starts by defining what your market really includes. Which demands do you satisfy? Which solutions can your customers choose between? Which customers have the realistic option to choose your product offering? You should then list the most important factors that have impact on your company, starting from overall factors and moving down to relevant micro factors. How does the eco-system look? You should work out which development stage your business is in and what that means to you. Should the company aim for overall consolidation, penetration, market development, product development or diversification?
2.2 Competition	The starting point may be a list of the main competitors and how they are doing. You may include a prediction of what the strongest competitors may be expected to do within the planning period, and list how you could react. It is important to consider positive as well as defensive alternatives. In reality, you should try to 'choose' who you decide to treat as competitors and who as partners.
3. Key competencies	List your key competencies and what you must do to ensure a lasting strength in the chosen markets. You may find it useful to imagine how you could justify an investment in your company to an external potential investor. Why not invest in the competitor instead? Why or how is your company stronger?
4. Main strategies	Here you list the conclusions you can draw from the preceding analysis. This has to be fairly short, as you have already provided the rationale. You should include: • Value model • Product portfolio • Segmentation strategy

Section	Examples of what may be included
	▪ Alliance strategy ▪ Network creation strategy
5. Marketing	As you have already listed the main strategies, it is now time to proceed to the specific strategies, each of which you should describe with a brief rationale
5.1 Product portfolio	Summarize here the products you will offer and the benefits that these offerings will bring to partners and end users
5.2 Price	Explain the pricing policies
5.3 Distribution	Describe your use of electronic and physical different distribution channels, how these are managed, how conflicts can be minimized, etc.
5.4 Promotion	This analysis will typically start with a description of the company's stakeholders and an evaluation of the overall purpose of the promotion activities (create attention, build image, etc.). Following that, describe the overall message. Finally, describe strategies for use of PR, advertising and promotion
6. Alliances	This is an examination of the companies or groups of companies that you should have alliances with. The key alliance concepts should be listed (see Appendix A)
7. Organization	The company's organization is often complex, with one structure based on activities and others based on line functions. You should describe main elements of the organization and how it should evolve during the planning period
8. Quantitative plan	This is a quantitative overview of sales, revenues, investments, profit contributions, etc.
9. Follow-up	The organizational responsibilities of implementing the plan are listed here. This will typically be a combination of routines for quantitative follow-up (how are your market, sales, quality measures, profitability etc. developing?), performance reporting and qualitative follow-up (are your strategies still correct?). It should also include what you believe are your best measurements of success. These may include financial measures, but also website page views and click-throughs, customer complaints, delivery times, service ratings, image ratings, percentage of goods returned, etc. These measures must reflect the overall analysis of your company's situation and you should set targets for them

The long-term marketing plan is only complete if it reflects a thorough marketing audit. This audit can be performed in a formal and structured way, or it can be more informal.

> ### Planning system or planning culture?
>
> Small, entrepreneurial companies in dynamic environments may have difficulties in implementing a formal strategic marketing planning system if dramatic changes in their sector environment and new opportunities change the picture very often. However, this should not prevent them from implementing a strategic planning culture. This culture may include frequent discussions of well-written strategic positioning documents.

SHORT-TERM MARKETING PLANNING

Writing Short-term Marketing Plans

If a company makes long-term marketing plans, then it is not necessary to repeat the same arguments in the short-term plans. However, the short-term plans need to be more specific about actions than the long-term plans. It is not possible to provide a standard model for what a short-term plan must include, but the template shown in Table 5.22 can serve as an inspiration.

Short-term Marketing Control

The purpose of the short-term marketing control process is to find out whether things work as planned, so that action can be taken whenever something goes wrong. It addresses the following basic questions:

- Do the market and the competition behave as planned?
- Does our promotion work as planned?
- Does our pricing work as planned?
- Does our sales and distribution work as planned?
- Do our products and services work as planned?
- Do we have the planned marketing profitability?

Table 5.22   Structuring the short-term marketing plan

1. Executive summary	Typically one or two pages that summarize what you expect to achieve within the planning horizon (which will typically be one year). In particular, you should emphasize milestones to be achieved. Remember that in contrast to an analysis, the short-term plan is not necessarily an attempt to answer specific questions, and it may consequently be meaningless to attempt to provide a 'conclusion'. A summary is always relevant, however
2. Analysis of business environment	The purpose is to describe what will happen in the marketplace within the planning period. This should, of course, include only factors that will affect the company directly. Inspiration may come from the long-term marketing plan
2.1 The market	Define the market that you compete in and the factors in the market that may affect the company. As the planning horizon of this plan is short, you should focus on factors that will have real impact within the planning period. This is most often factors on the micro rather than the macro level, and may include anticipated events relating to specific customers and orders (for capital goods), exhibitions, etc.
2.2 Competition and alliances	List your most important competitors within the planning period and provide an overview of their product offering, communication activities, pricing, service patterns, distribution networks, etc. You may then list your responses, such as attempts to block distribution, match prices, advertise, engage in exclusive deals, etc. If you aim to make alliances, then you should also list that here
3. Milestones	Based on section 2, you can now list some milestones for the period. These should be simple to understand and communicate, and may include improvements of quality, service, production, brand awareness and, of course, sales and profitability. You will typically want to define deadlines, quantitative measures, and responsibilities, and you may define rewards for meeting the milestones
4. Marketing	As milestones have now been defined, you should establish what needs to be done to meet them. This should be considerably more concrete than the corresponding sections in the long-term plan. The treatment will typically be divided according to your market segments and, if your company operates internationally, will be divided into geographical areas

(continued over)

Table 5.22 *continued*

4.1 Product portfolio	This is a description of the products that you will offer for each market and channel, in what quality, with which basic specifications, etc.
4.2 Service	Describe services as stand-alone offerings or as parts of your product offerings. You may include elements such as electronic one-to-one marketing and product-related chat forums
4.3 Price	This is a listing of prices and price policies. Responsibilities should be clearly defined
4.4 Promotion	List promotion activities, perhaps month by month. The list may include exhibitions and fairs, Web branding, roadshows, etc.
4.5 Distribution	This section should describe your channel structure, sales targets, sales commission structure, etc.
4.6 Logistics	Logistics is a central issue in many companies. The description may include how you will handle electronic order processing and challenges for complex or very large orders. It will often be based on what happens from the initial contact with a potential customer to the delivery and payment
5. Quantitative plan	This will typically include: • Revenues • Profit • Contributions • Sales-related costs • Efficiency/quality measures Companies with mass marketing activities will often have a measuring system for the sales chain that includes measurements of the entire electronic click-stream, including what people purchased, how many returned the products, etc. This may be broken down into districts and sales people, etc. Companies with large ad hoc sales transactions will typically make a presales calculation and an after-sales calculation to learn about sources of deviations and misconceptions.
6. Follow-up	This is a specification of which factors and variables you need to measure during the planning period

Marketing control is typically based on a combination of field research, desk research and online marketing research methods (Figure 5.16). The more connected the company is with its distribution and end users, the more important online marketing research methods are.

Idea phases
- Who should we target?
- What benefits should we deliver and at what prices?
- How should we get the product to the target groups?
- How should we make the target groups aware of the benefits that we can deliver?

Control phases
- Do the market and the competition behave as planned?
- Does our promotion work as planned?
- Does our pricing work as planned?
- Does our sales and distribution work as planned?
- Do our products work as planned?
- Do we have the planned marketing profitability?

Online marketing research generates real-time one-to-one information about the company's marketing effort

Development phases
- How do we ensure that the end users will accept our concept?
- How do we create products that fully satisfy the demand?
- How do we ensure that our communication activities will work?
- How do we ensure that the product and communication are consistent?
- How do we ensure that product packaging is accepted?

Figure 5.16   Using online research to gather marketing information. Information about a company's marketing activities can be divided into the information it needs at each stage of the planning cycle for a given activity.

Appendix E contains an overview of typical control factors and the data input that can be used as a basis for each.

THE ESSENCE

'In skating over thin ice, our safety is in our speed.'

Ralph Waldo Emerson

THE ESSENCE

CHAPTER 6

THE ESSENCE

We have now looked at nearly all the major aspects of the digital school, so perhaps it is time to reflect on what the essence of it all is. Our first observation is that what is happening to the business of marketing is indeed a big change. Any one who can't see this should think about where Microsoft, Cisco, Yahoo!, Nokia and AOL came from. Four of these companies weren't even around a few years ago, and while the fifth did exist, it wasn't terribly sexy. Nokia's claim to fame was, until quite recently, as the manufacturer of rubber boots. If this is not enough to impress, then consider what it will be like when computers are a thousand times faster than today, so that they can drive cars and cut our grass. And what will it be like when we all have 200 devices connected to online networks. It *is* big but it has only just begun. Trust us on this one.

But what are the essential implications for marketing? We have concluded that four are particularly important:

- network effects
- increased use of leverage

- minimal marginal costs
- path dependency.

These characteristics are interlocked, but if some of the things that people do in marketing today seem strange, then the explanation might lie in the way that these phenomena now work. Take, for instance, leverage. The use of leverage by successful companies in the new economy is indeed extreme. They issue millions of shares to raise money so that they can grow into the increasing returns territory. They do it not necessarily because they are irresponsible, but because it's often the only way. And then they leverage themselves further in the way that they cooperate. They crawl all over each other to exchange branding, standards, production facilities and marketing vehicles to get to where they want to go. They all do it – every winning company – because each of them has a mission that is far greater than what they could ever achieve through use of their own resources only. It is about leveraging yourself through the work of others – which is fine, because the others do it to (or with) you as well.

Most don't make it, but some do, and in a big way. Those that make it reach a (temporary) Nirvana where they enjoy effects that are truly new. They enjoy network effects where the value of what they provide to each customer increases as more and more customers switch on, and where the returns on investments just keep going up. Not forever, of course, but they keep on rising further than companies in traditional industries would experience.

The digital school of marketing describes an adrenaline-driven race to a leadership that starts feeding on itself as network effects and standards leadership create lock-in and switching costs for the customers. Reduced down to a single word: *speed*.

The Americans actually have a joke about this. Two men meet a big bear in the forest. One of them sits down and starts to put on his running shoes. The other looks at him and says, 'It's no use. You can't outrun a bear anyway'.

The first one answers, 'I don't have to outrun the bear. I just have to outrun you.'

The new economy is here and it has just begun. Enjoy the race.

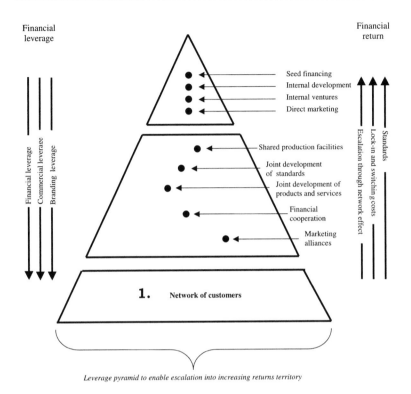

Leverage pyramid to enable escalation into increasing returns territory

Figure 6.1 Marketing in the new economy. Businesses operating in the new economy need to move very quickly to the stable situation where they benefit from network effects. The road towards that situation is largely based on numerous partnerships with other companies in the ecosystem. Each company in these partnerships leverage their resources to achieve the speed and impact necessary to achieve leadership.

COOPERATION POSSIBILITIES IN A NETWORKED ECOSYSTEM

DEVELOPMENT OF STANDARDS

Cooperation approach	Explanation	Advantages	Disadvantages
Participation in open standards forum	Active participation in the development of open standards	Ability to steer standards towards areas where the company is strong, and to gain control over its future development	Consumes considerable time and resources The process can be very slow and the agreed direction may not be in the interest of the company
Participation in proprietary standard consortium	Active participation in the development of proprietary standards	Ability to steer standards towards areas where the company is strong, and to gain control over its future development	Consumes considerable time and resources May lose out to open standards

PRODUCTION FACILITIES

Cooperation approach	Explanation	Advantages	Disadvantages
Sharing of logistic facilities	Companies share shipping, storage, billing, computing or other facilities	Cost savings Rationalization More efficient operation	Coordination problems
Joint production facilities	Companies share factory facilities	Cost savings Rationalization Joint development of know-how	Coordination problems
Resource exchange	Exchange of staff, and physical resources, etc. This may be in the context of export cartels or other groups	Cost savings Exchange of know-how	Coordination problems
Know-how exchange	Companies form discussion forums, such as export groups, to exchange know-how about markets, technologies and processes	Effective access to know-how May stimulate other cooperation	Secrecy issues

DEVELOPMENT OF PRODUCTS AND SERVICES

Cooperation approach	Explanation	Advantages	Disadvantages
Acquire licence	Take a licence to use technology and/or trademark from existing supplier (which will	Fast Inexpensive in the short run May give rapid market strength that	Dependence Limited access to know-how Licence fee Licence provider

Cooperation approach	Explanation	Advantages	Disadvantages
	typically have a head start)	could otherwise be difficult to get	may lose its competitive strength
Sell licence	Company has a trademark or technology that can be licensed to another company (which will often have more sales power)	Fast Very low risk Licence revenues	Licence taker may block company's own expansion later on

FINANCIAL COOPERATION

Cooperation approach	Explanation	Advantages	Disadvantages
Market-oriented acquisition	Buy a company or part of a company that has a competitive advantage in the market	Fast Access to know-how, trademarks, distribution network, customers, etc.	Expensive Risk of loss of key staff Integration problems
Know-how-oriented acquisition	Buy a company that has know-how advantages with the aim to transfer these advantages to your existing organization	Fast Access to know-how, human resources and key technologies	Expensive Risk of loss of key staff (may view you as an intellectual 'blood sucker') Integration problems
Network-based joint venture	Establishment of a new unit/company in cooperation with partner(s) to secure joint market leadership for network effect	Fast Inexpensive May provide synergism	Management conflicts May block potential of approaching the market alone later on

Cooperation approach	Explanation	Advantages	Disadvantages
Cost-sharing-based joint venture	Establishment of a new unit/company in cooperation with partner(s) to share high start-up costs	Fast Inexpensive May provide synergism	Management conflicts May block potential of approaching the market alone later on
Complementary strength-based joint venture	Establishment of a new unit/company in cooperation with partner(s) to combine critical skills and assets, such as media assets and technology assets	Fast Inexpensive May provide synergism	Management conflicts May block potential of approaching the market alone later on
Venture capital investment	Invest in new or smaller company	May provide understanding of the technology and the market Provides corporate insight at limited cost May be a precursor to further investment in the company	Gives only limited market access Gives only limited access to know-how Gives very limited control

MARKETING ALLIANCES

Cooperation approach	Explanation	Advantages	Disadvantages
Cross sales	Two companies refer customers to each other	Increased sales Inspiration Partly neutralizing potential competitor Can enhance image with customer	Risk of helping potential future competitor

Cooperation approach	Explanation	Advantages	Disadvantages
Joint promotion	Cobranding, joint exhibition activities, joint PR activities, etc.	Cost savings Inspiration Partly neutralizing potential competitor Enhanced image	Confusion from customer perspective Conflicts of interest Coordination problems
Joint pricing	Offer the customers advantages if they buy combination of products from the two cooperating companies	Increased sales Enhanced image	Customer confusion Conflicts of interest Coordination problems
Integrated sales	Sales of combined products containing elements from both companies	Enhanced total product offering Partly neutralizing potential competitor Enhanced image	Customer confusion Conflicts of interest Coordination problems
Joint promotion organization	Several companies join forces to form jointly owned promotion organization	Cost savings Authoritative image	Customer confusion Conflicts of interest Coordination problems
Joint promotion facilities/staff	Companies share sales offices, sales staff, exhibition resources, etc.	Cost savings Increased resources	Customer confusion Conflicts of interest Coordination problems
Joint branding	A franchise-inspired solution, in which several independent companies share a brand	Critical mass in image building Savings	Customer confusion Conflicts of interest Coordination problems
Sharing of sales administration	Companies share sales administration and may invoice jointly	Cost savings Rationalization Advantage for end user	Coordination problems, in particular in relation to special offers

END-USER SEGMENTATION CRITERIA

	Business-to-consumer markets	Business-to-business markets
Geography	Country Region	Country Region
Demographics	Economic status Culture/ethnic groups Political orientation	Economic status Culture/ethnic areas Political orientation
Social/sector criteria	Education level Profession Household income Lifestyle group	Corporate professional level Business sector Size of the company/organization Technological development Production facilities Financial status
Personal criteria	Sex Age Marital status Size of household (number of people)	Decision maker's position in the company Decision maker's age Decision maker's sex
Price sensitivity	Price sensitivity	Price sensitivity

	Business-to-consumer markets	Business-to-business markets
Purchasing situation	Usage intensity Importance of the product to the user Brand loyalty Geographical purchase preferences Selection criteria in the purchase situation Distribution channel	Size and frequency of ordering Importance of the purchase Brand loyalty Geographical purchase preferences Selection criteria in the purchase situation Distribution channel Need for partial or total turnkey solutions
Product usage	How is the product used? What features in the product are most appreciated? What is the real reason for the purchase? Service requirements Substitution options	How is the product used? What features in the products are most appreciated? Service requirements Substitution possibilities
Benefit	Standard benefits Corporate benefits Differentiation benefits Network benefits	Standard benefits Corporate benefits Differentiation benefits Network benefits

IMPORTANT FIELD RESEARCH METHODS

Research method	Scenario	Typical time requirements
Focus groups	You do not yet have a structured view of the respondents' attitudes to your (still) loosely formulated ideas about potential products, advertisements or packaging concepts. You may be looking to create the basis for later quantitative tests. The respondents are specialists or the subjects to discuss are sensitive/personal	About 4 weeks
Informal interviews	Same as for focus groups	1–4 weeks
Key interviews	You need information about market structures and competitors, and you need qualified opinions and advice in addition to facts	Varies
Distribution research	You need to evaluate relevant opinions and attitudes of executives in the distribution chain as well as your competitors' and your own company images, competitiveness and product flow in the distribution channels	Varies

Research method	Scenario	Typical time requirements
Retail panels	You want to track how specific retail units are purchasing, storing, selling and distributing your own and competing brands	Subscription-based
Radar	Mapping of product target groups and their attitudes towards your product	1–4 weeks
Supply mapping	You need to track products that are available within a sector of the market. This may include types of product, quantities, prices, presentation and packaging, and shelf space allocation to them. The result can be an impression of distribution patterns, competition, price structures, and the parameters necessary to succeed in the market	Few days
Consumer panels	You are interested in tracking how end users are using your own and competing brands, and how that will evolve over time	Varies
Concept tests	You need to evaluate product concepts to find out whether they will work, what their strongest and weakest elements are and, if there are several alternatives, which one is best. The test may be conducted separately for products and packaging	About 4 weeks
Product tests	You need to evaluate product features to find out whether they will work, what their strongest and weakest elements are and, if there are several alternatives, which one is best. The test may be conducted separately for products and for packaging. The test is performed on specific features	About 6 weeks
Advertisement pretest	Evaluation of alternative advertisements to determine whether they will work satisfactorily, what their strongest and weakest points are and, if there are several alternatives, which one is best	About 4 weeks

Research method	Scenario	Typical time requirements
Packaging pretests	Evaluation of alternative packaging solution's ability to communicate the product idea	About 4 weeks
Concept/ product test	You need to evaluate product concepts and products to evaluate the overall impression and possible discrepancies between concept and product	About 8 weeks
Market maps	You are selling a mass marketing brand and you need a quantitative overview of your product's position versus competing products. You are also interested in identifying product/marketing opportunities for your own and competing products	Subscription based

INPUT FROM THE MARKETING AUDIT TO THE STRATEGIC MARKETING PLAN

	Defining corporate mission statement	Value audit	Economic environment audit	Business ecosystem audit	Competitor audit	End–user audit	Product positioning audit	Pricing strategy audit	Distribution strategy audit	Promotion strategy audit	Speed and timing audit	Concluding from the audit
1. Summary	X	X	X	X	X	X	X	X	X	X	X	X
2. Analysis of the business environment			X	X			X				X	X
2.1 The market				X	X	X	X				X	X
2.2 Competition					X		X				X	X
3. Key competencies											X	X
4. Main strategies		X					X	X	X	X	X	X

	Defining corporate mission statement	Value audit	Economic environment audit	Business ecosystem audit	Competitor audit	End-user audit	Product positioning audit	Pricing strategy audit	Distribution strategy audit	Promotion strategy audit	Speed and timing audit	Concluding from the audit
5. Marketing												
5.1 Product portfolio package							X				X	X
5.2 Price								X			X	X
5.3 Distribution									X		X	X
5.4 Promotion										X	X	X
6. Alliances	X						X		X		X	X
7. Organization									X		X	X
8. Quantitative plan					X	X	X	X	X	X	X	X
9. Follow-up					X	X	X	X	X	X	X	X

MARKETING CONTROL METHODS

Question	Operational control indicators (actual versus planned situation)	Commonly used information sources
Do the market and the competition behave as planned?	Controlling market growth and behaviour	Email interviews Online votes Wish lists and shopping baskets Cyber-panels Internet search Statistical publications Press-scanning services Periodicals Multiclient studies Online databases
	Controlling competitor behaviour	Email paging Email interviews Cyber-panels Online focus polls Internet search Statistical publications Press-scanning services

Question	Operational control indicators (actual versus planned situation)	Commonly used information sources
		Periodicals Multiclient studies On-line databases
Does our promotion work as planned?	Controlling development in relative brand awareness, usage and preference for the company and its competitors among end users	Click-stream and page-view tracking Email interviews Online votes Cyber-panels Online focus polls Market maps
	Controlling promotion efficiency for each medium used	Click-stream and page-view tracking Link tracking Email interviews Wish lists and shopping baskets Cyber-panels Online focus polls Market maps Retail panels Consumer panels Distribution research
	Controlling image and market position in distribution and among end users	Email interviews Cyber-panels Virtual pretesting Online focus polls Consumer panels Distribution research Focus groups Informal interviews
Does our pricing work as planned?	Controlling pricing levels through the distribution chain	Internal pricing reports Distribution research Supply mapping Trade statistics

Question	Operational control indicators (actual versus planned situation)	Commonly used information sources
	Controlling competitive price levels at end-user level	Internal pricing reports Consumer panels Distribution research Retail panels Supply mapping
	Controlling discount levels and discounting frequency	Click-stream and page-view tracking
	Controlling sales funnel, including initial contact, enquiries, sales meetings, orders and return goods/lost accounts (and ratios between these)	Click-stream and page-view tracking Link tracking Focus groups
Does our sales and distribution work as planned?	Controlling sales per unit, team and salesperson	Click-stream and page-view tracking One-to-one marketing
	Controlling sales volumes and market share developments per distribution network segment	Click-stream and page-view tracking Distribution research Supply mapping Trade statistics
	Controlling sales volumes and market share developments per end-user segment	Click-stream and page-view tracking Distribution research Supply mapping Trade statistics
	Controlling distributor's satisfaction with products and services	Email interviews Cyber-panels Online focus polls Radars Distribution research Focus groups Informal interviews

Question	Operational control indicators (actual versus planned situation)	Commonly used information sources
Do our products and services work as planned?	Controlling end users' satisfaction with products and services (opinion stated, resales and reference sales ratios, returned goods ratios, complaints, etc.)	Click-stream and page-view tracking Email interviews Online votes Wish lists and shopping baskets Cyber-panels Focus polls Radars Focus groups Market maps
	Controlling percentage delivered on time	Click-stream and page-view tracking Radars
	Controlling product/service quality (defects, returned goods, warranty expenses, etc.)	Click-stream and page-view tracking Email interviews Radars Distribution research Focus groups Informal interviews
Do we have the planned marketing profitability?	Controlling cost and margin per contact and order	After-sales calculations and other internal reports
	Controlling cost and margin per unit, team and salesperson	Internal reports
	Controlling cost and margin per distribution and end-user segment	Internal reports
	Controlling cost and margin per product category	Internal reports

INFORMATION INPUT FOR THE MARKETING AUDIT

	Defining corporate mission statement	Value audit	Economic environment audit	Business ecosystem audit	Competitor audit	End-user audit	Product positioning audit	Pricing strategy audit	Distribution strategy audit	Promotion audit	Speed and timing audit
Information intermediaries											
Company directories					X				X		
VIP directories					X				X		
Research bureaus				X	X	X		X			
Credit rating agencies	X								X		
Email directories	X	X							X		

	Defining corporate mission statement	Value audit	Economic environment audit	Business ecosystem audit	Competitor audit	End-user audit	Product positioning audit	Pricing strategy audit	Distribution strategy audit	Promotion audit	Speed and timing audit
Public institutions			X	X	X				X		
Trade associations			X	X	X				X		X
Embassies			X	X	X				X		
Financial institutions			X		X						X
Information consultants			X	X	X	X		X	X		X
Secondary sources											
Statistical sources					X	X	X			X	X
Press-scanning services		X			X	X	X		X	X	X
Periodicals		X			X	X	X		X	X	X
Multiclient studies		X			X	X	X		X	X	X
Online databases		X			X	X	X		X	X	X
Primary sources											
Competitors			X	X					X	X	X
Subcontractors				X				X		X	X
Wholesale			X	X				X	X	X	X
Retail			X						X	X	X
End users						X			X	X	X
Field research methods											
Focus groups		X	X	X	X			X	X	X	X
Informal interviews		X	X	X				X	X	X	X
Key interviews		X	X	X				X	X	X	X

	Defining corporate mission statement	Value audit	Economic environment audit	Business ecosystem audit	Competitor audit	End-user audit	Product positioning audit	Pricing strategy audit	Distribution strategy audit	Promotion audit	Speed and timing audit
Distribution research			X	X	X			X	X	X	X
Retail panels			X	X	X	X		X	X	X	X
Radar			X	X	X			X	X	X	
Supply mapping			X	X	X	X		X	X	X	
Consumer panels				X	X	X		X	X	X	
Concept tests						X		X	X	X	
Product tests						X		X	X	X	
Advertisement test						X		X	X	X	
Packaging tests						X		X	X	X	
Concept/product tests						X		X	X	X	
Market maps			X	X	X			X	X	X	X

Online market research

	Defining corporate mission statement	Value audit	Economic environment audit	Business ecosystem audit	Competitor audit	End-user audit	Product positioning audit	Pricing strategy audit	Distribution strategy audit	Promotion audit	Speed and timing audit
Click-stream and page-view tracking						X					
Link tracking									X		
Email paging							X			X	
Email interviews						X	X	X	X	X	
Online votes						X	X				
Wish lists and shopping baskets						X	X	X		X	
Cyber-panels					X	X	X	X	X	X	
Virtual pretesting							X				
Focus polls						X	X	X	X	X	
One-to-one marketing						X	X	X	X	X	

LEADING MARKETING SCHOOLS

THE COMMODITY SCHOOL

Brief description
Focuses on the physical characteristics of products and services and on how customers respond to these characteristics.

Example
For product X, what is the typical customer's brand insistence, average purchase quantity, ego-involvement, and shopping effort?

The school emerged in the early 1910s and has been very influential.

Key publications

- Charles Parlin (see Gardner 1945) was the initial proponent of the commodity school.
- Copeland (1923) provided pioneering product classification system (convenience, shopping and speciality goods).

- Holbrook and Howard (1977) suggested a modified classification, which they tied to clarity, self-confidence, mental shopping effort, brand insistence, purchase magnitude, ego-involvement and physical shopping effort.

THE FUNCTIONAL SCHOOL

Brief description
Listing and classification of the tasks involved in marketing practice.

Example
What are the main tactical tasks to be performed for international marketing of product X, and how should we organize their execution?

The functional school has developed little on a theoretical basis since the 1950s, but it has been very influential in terms of clarifying marketing roles and planning tasks.

Key publications

- Shaw (1912) divided marketing functions into sharing the risk, transporting the goods, financing the operations, selling, ad assembling, assorting and reshipping.
- McGarry and Edmund (1950) were among a group of subsequent writers who improved this classification. McGarry divided the functions into the contractual, merchandising, pricing, propaganda, physical distribution and termination functions.
- McCarthy (1960) popularized 'the four Ps' of marketing (product, price, promotion and place).
- Porter (1980) popularized the concept of the value chain and explained how it could be used to clarify the desired role of the company in the marketplace and in its relationship with competitors, suppliers, distributors and alliance partners.

THE REGIONAL SCHOOL

Brief description

Focuses on how to bridge the physical gap between buyers and sellers.

Example

How do we optimize location and size of wholesale and retail outlets for our products?

The regional school has high relevance for several specialized marketing disciplines.

Key publications

- Nystrom (1913) published a classical book about retailing. In this and subsequent writings, he described the roles and functions of different retaining organizations, theories of location, rent, price maintenance, etc.
- Converse (1949) modified the models and tested them empirically.
- Revzan (1961) proposed models for calculating ideal wholesale market areas. The models were based on product weight relative to value, relative perish ability, product differentiating techniques, factors affecting plant location, price and price policies, transportation rates and services, individual firms' marketing methods and auxiliary services.
- Huff (1964) proposed models for calculating the ideal size of shopping centres.
- Grether (1950) generalized many of the regional school's thoughts and integrated them into broader marketing frameworks.

THE INSTITUTIONAL SCHOOL

Brief description

Focuses on the organizations required to perform marketing tasks successfully.

Example

How do we organize ourselves and work with other companies and institutions in order to optimize the execution of our marketing tasks?

This school emerged in the 1910s, largely as a response to consumers' lack of understanding of the gaps between farmers' sales prices and retail prices for agricultural products. Its concepts clarified how a corporation or institution should organize itself for marketing success. An example of early adoption was Procter & Gamble, which was probably the first company to designate the title 'Marketing Manager' to an employee.

Key publications

- Weld (1916) was a leading pioneer who described the efficiency and necessity of the marketing channels.
- Butler (1923) described the role of middlemen in the distribution chain.
- Breyer (1934) delivered a well-argued rationale for the need for marketing institutions.
- Converse and Huegy (1940) described advantages and disadvantages of vertical integration.
- Alderson (1957) pioneered the development of theories about marketing management, as he explained executive marketing management processes drawing upon concepts from psychology, sociology, anthropology and general economics.
- McCammon (1963) suggested using concepts from psychology and sociology to explain why marketing organizations do not always evolve optimally. In 1965, he provided a clear explanation of the advantages of integrated marketing strategies and organizations.

THE MANAGERIAL SCHOOL

Brief description
The study of how marketing concepts are changed into tasks, and how these tasks are managed.

Example
How do we put together a complete plan for penetrating market X?

This school is part of the managerial economics movement that took off in the late 1940s and early 1950s. Consultants and corporations, such as Procter & Gamble, General Foods, General Electric, J. Walter Thompson, and Boston Consulting Group, and the academic community have driven the development of this school of thought. Many of the concepts were consciously applied in practice long before the models were published.

Key publications

- Scott (1903), Director of Psychological Studies at Northwestern University, concluded that psychology was the only discipline that could be used as a basis for theories about advertising. He built advertising theory incorporating such concepts from psychology as perception, illusion, suggestion, association, attention and imaginary.
- Hollingworth (1913) described psychological principles for receiving responses to advertising, including measurement of appeals and memory recall.
- One of the first attempts to formulate an integrated, multidisciplinary theory of advertising was given in 1915 by Tipper (an advertising manager), Hollingworth (a psychologist), Hotchkiss (a business English professor) and Parsons (a professor of fine and applied art).
- Shaw (1916) dedicated a chapter to market analysis in a book about marketing. He divided markets into strata and suggested that these should be analysed individually.

- Adams (1916) was one of the first to suggest practical methods for testing advertisements with mathematical assurance. He suggested that ads could be broken into their elements, which could be analysed separately.
- White (1921) argued that markets are measurable and that the analysis of them can provide guidance for production, sales, development and promotion.
- Russel (1924) wrote one of the first books with an integrated description of salesmanship.
- Reilly (1929) described techniques for questionnaire-based field studies, explaining that it would be possible to use distributors, consumers, jobbers and institutions as sources of market information.
- White (1929) described how to define sales targets and to develop market forecasts and sales forecasts.
- White (1931) described in detail how to prepare and execute questionnaire-based field studies. He included telephone interviews and focus groups.
- Kleppner (1933) was among the first to attempt to relate advertising strategies to product lifecycles. He distinguished between advertising in the pioneering stage (to open new markets), the competitive stage (to strengthen competitive position), and the retentive stage (to guard position).
- Wheeler (1937) published an overview of how questionnaires could be used in market research.
- Brown (1937) described quantitative market research, sales analysis, market trends, advertisement tests and product tests. He suggested that the ideal market analyst should combine knowledge about statistics psychology, accounting, engineering, sociology and marketing.
- Ferber (1949) described how to use sampling techniques, multivariate analysis and correlation analysis in market research.
- Dean (1950) introduced a number of new concepts for pricing policies, including 'skimming' and 'penetration' pricing.

- Dean (1951) wrote an influential textbook where he focused on showing 'how economic analysis can be used in formulating business policies'.
- Smith (1956) introduced the concept of marketing segmentation.
- Howard (1957) and Kelly & Lazer (1958) and other authors published a number of popular textbooks.
- McKitterick (1957) was perhaps the first to state clearly the modern 'marketing concept', in which the focus of the company is to identify and satisfy the demands of the customers.
- The American Management Association (1959) published a number of papers about marketing audits.
- Levitt (1960) published 'Marketing myopia', which is probably one of the most famous marketing articles of all times. The article explained how companies and sectors could lose their edge because they erroneously defined their mission as to deliver a given product rather than to satisfy a given demand.
- McCarthy (1960), Borden (1964) and Levitt (1965) were among the first to describe the implications of the product lifecycle.
- Magee (1960) and Davidson (1961) contributed to the understanding of the importance and choices regarding distribution strategies.
- Oxenfeldt (1960) proposed a multistage pricing approach, in which a company would first select target markets, then choose brand strategy, compose a marketing mix, select price policy, determine pricing strategy, and finally set the actual price.
- Lavidge & Steiner (1961) suggested that advertising should move potential customers through successive stages, beginning with creating fundamental awareness, creating favourable attitudes, creating preference, creating desire to buy, convincing and finally translating to actual purchase.
- Drucker (1964) pointed out the importance of making corporate strategy statements both specific and precise.
- Ansoff (1965) pioneered the concept of strategic planning in the corporation and introduced such tools as the Product/Mission Matrix and the Diversification Matrix.

- Buzzotta et al. (1972) published a classical book about the use of psychology in the sales process.
- Hedley (1977) described the Boston Consulting Group Matrix.
- Ansoff (1980) was one of the pioneers of the concepts of strategic marketing management.
- Reibstein & Gatignon (1984) and Petroshius & Monroe (1987) described product line pricing.
- Huber, Holbrook & Kahn (1986) contributed an important article about price sensitivity.

THE BUYER BEHAVIOUR SCHOOL

Brief description
Focuses on the reasons why people buy given products and services.

Example
What are the exact reasons that customers in segment X prefer to buy product Y?

This school, which emerged in the early 1950s, has adopted a range of concepts and methodologies from sociology, psychology and anthropology to go beyond assumed rational behaviour and understand the real reasons for buyer behaviour.

Key publications

- Katona (1953) emerged as one of several pioneers of the school, when he outlined the differences between economic and psychological consumer behaviour.
- Dichter (1947, 1964) pioneered motivation research, which is a part of the basis for focus groups and informal interviews.
- Katz & Lazarsfeld (1955) contributed significant input regarding opinion leadership and buyer behaviour.
- Festinger (1957) introduced cognitive dissonance to the buyer behaviour framework.

- Bourne (1957) described how a buyer can be influenced by reference groups.
- Rogers (1962) became famous for his analysis of diffusion of innovations (brands and products) among consumers.
- Kuehn (1962) inspired an avalanche of research into brand-buying behaviour.
- Holloway (1967) was one of the pioneers of research and field experimentation in cognitive dissonance theory and brand choice behaviour.
- Robinson, Faris & Wind (1967) published some of the early studies of industrial buying behaviour.
- Engel, Kollat & Blackwell (1968) provided a model for consumer behaviour and variables that can shape consumer action.
- Sheth (1974) contributed some of the first studies of collective buying decision-making within families.
- Howard & Sheth (1969) pioneered the research into the relationship between attitudes and buying behaviour.

THE DIGITAL SCHOOL

Brief description
Focuses on how marketing is exercised in economic sectors characterized by widespread use of digital networking.

Example
Which marketing strategies are specifically relevant for companies whose business is strongly dominated by digital data processing and networking?

The school gained ground from the mid 1980s and reached mainstream as telecommunications, media and technology sectors boomed in the 1990s.

Key publications and events

- Leibenstein (1950) published an article that described the 'bandwagon, snob, and Veblen effects'. This article is largely seen as a precursor for the network effect literature that developed after the launch of digital networking. His main claim was that demand curves are more elastic when consumers derive positive value from increases in the size of the market.
- Moore (1965) suggested in an article that chips' processing power will continue to grow exponentially for a sustained period of time. This was developed later into what is now called 'Moore's Law'.
- Todler (1970) introduced the term 'prosumer' to describe the consumer who is involved in the production process.
- David (1985) provided examples of how inferior technologies may succeed because of network effects.
- Katz and Shapiro (1985) published a very influential paper that defined network effects: 'There are many products for which the utility that a user derives from consumption of the good increases with the number of other agents consuming the good.'
- Weiser (1988) coined the term 'ubiquitous computing'.
- Arthur (1989) provided a key overview of how increasing returns and lock-in could be key factors in the development of high-tech markets. He considered path dependence as a 'lock-in by historical events,' which was likely when there were increasing returns.
- Arthur (1990) published an article highlighting the increasing number of positive feedback processes in the economy.
- Davis and Davidson (1991) were among the first to prove the concept of a market segment of one.
- Geoffrey Moore (1991) described the difficulty of moving from early-adopter markets to mainstream markets.
- Gilder (1993) coined the term 'Metcalf's Law', after George Metcalf, who invented the Ethernet and founded 3Com Corp. The

law states that the value of a network user is proportional to the square of the number of users.

- Drucker (1993) argued that in the new economy, the key means of production was no longer capital equipment but human intellect. This shifted the power from corporations to knowledge workers.
- Besen and Farrel (1994) provided an overview of how network externalities could be applied in the choice and development of standards.
- Gilder (1994) writes the article 'The Bandwidth Tidal Wave' in *Forbes ASAP*, where he suggests that bandwidth will grow much faster than chip processing power and that networks will become increasingly central to computing.
- Liebowitz and Margolis (1995) discussed whether network effects could be characterized as market failures.
- McKnight (1995) proposed that the Internet could best be understood as operating on the principles of its positive network externalities, statistical sharing and interoperability.
- Economides and Himmelberg (1995) published a classical analysis of network effects using the US fax market as an example. They concluded that perfect competition does not generate the optimal outcome in a market dominated by network effects.
- Kauffman (1995) showed that as the number of connections among elements in a system rises above half the number of elements, the probability of cascading events rises dramatically. This represents a phase change in the system, resulting in non-linear responses to external inputs.
- Moore (1995) coined the term 'business ecosystems'.
- Arthur (1996) provided an overview of how network effects were changing business environments.
- Hagel and Armstrong (1997) suggested that technology should not only help customers generate information about users, but also help users to generate information about themselves and about economies.
- Krugman (1997) stated that in the network economy, supply curves slope down instead of up, and demand curves slope up instead of down.

- Saffo (1997) created the term 'disinter-remediation' to describe the situation where the new economy has changed the role of intermediaries but does not eliminate them. He also introduced the term 'value web' the same year.

SOURCES ON TRADITIONAL MARKETING SCHOOLS

The following list provides selected classical books and articles, as well as more recent literature, that have provided the basis for many of the traditional models and concepts described in this book.

Aaker, David A., 'Managing assets and skills: the key to a sustainable competitive advantage', *California Management Review*, Winter 1989, pp. 91–106.

Aaker, David A., *Managing Brand Equity*, New York, Free Press, 1991.

Abell, Derek F., Strategic Windows, *Journal of Marketing,* July 1978.

Adams, Henry F., *Advertising and its Mental Laws*, New York, Macmillan, 1916.

Alderson, W., *Marketing Behavior and Executive Action*, Homewood, Irwin, 1957.

American Management Association, *Analyzing and Improving Marketing Performance*, Report No. 32, 1959.

Ansoff, H. Igor, *Corporate Strategy*, New York, McGraw-Hill, 1965.

Ansoff, H. Igor, 'Strategic issue management', *Strategic Management Journal*, April–June 1980, pp. 131–148.

Bechmann, Theodore N., *Wholesaling*, New York, Ronald Press Co., 1926.

Blankenship, A.B., *Consumer and Opinion Research*, New York, Harper & Bros, 1943.

Bloom, Paul N. and Kotler, Philip, 'Strategies for high market share companies', *Harvard Business Review*, November–December 1975, pp. 62–72.

Bonoma, Thomas V. and Shapiro, Benson P., *Segmenting the Industrial Market*, Lexington, Lexington Books, 1983.

Borden, Neil H., 'The concept of the marketing mix', *Journal of Advertising Research*, June 1964, pp. 2–7.

Boston Consulting Group, 'The rule of three and four', *Perspectives*, 187, 1976.

Bourne, Francis S., 'Group influence in marketing and public relations', in Rensis Likert and Samuel P. Hayes, Jr, eds, *Some Applications of Behavioral Research*, Paris, United Nations Educational, Scientific and Cultural Organization, 1957, pp. 207–257.

Breyer, Ralph F., *The Marketing Institution*, New York, McGraw-Hill, 1934.

Brown, Lyndon O., *Market Research and Analysis*. New York, Ronald Press, 1937.

Butler, Ralph Starr, *Marketing and Merchandising*, New York, Alexander Hamilton Institute, 1923.

Buzzell, Robert D., ed., *Mathematical Models and Marketing Management*, Boston, Harvard University Press, 1964.

Buzzell, Robert D. and Gale, Bradley T., *The PIMS Principles*, New York, Free Press, 1987.

Buzzotta, V.R. et al., *Effective Selling Through Psychology*, New York, Wiley, 1972.

Converse, Paul D., 'New Laws of retail gravitation', *Journal of Marketing*, October 1949, p. 384.

Converse, Paul D. and Huegy, Harvey, *The Elements of Marketing*, New York, Prentice-Hall, 1940.

Copeland, Melvin T., 'The relation of consumers' buying habits to marketing methods', *Harvard Business Review*, April 1923, pp. 282–289.

Davidson, William R., 'Channels of distribution – one aspect of marketing strategy', *Business Horizons,* February 1961, pp. 84–90.

Day, George S., 'Diagnosing the product portfolio', *Journal of Marketing*, April 1997, pp. 29–38.

Dean, Joel, 'Pricing policies for new products', *Harvard Business Review*, November 1950, pp. 45–53.

Dean, Joel, *Managerial Economics*, Englewood Cliffs, Prentice-Hall, 1951.

Dichter, Ernest, *Handbook of Consumer Motivation: The Psychology of the World of Objects*, New York, McGraw-Hill, 1964.

Dichter, Ernest, 'Psychology in market research', *Harvard Business Review*, Summer 1947, pp. 432–443.

Dichter, Ernest, 'The world consumer', *Harvard Business Review*, July–August 1962, pp. 113–122.

Drucker, Peter, 'The big power of little ideas', *Harvard Business Review*, May–June 1964

Drucker, Peter, *Innovation and Entrepreneurship: Practice and Principles*, New York, Harper and Row, 1985.

Drucker, Peter, *Post-Capitalist Society*, New York, Harper Collins, 1993.

Engel, J.F., Kollat, D.T. and Blackwell, R.D., *Consumer Behaviour*, New York, Holt, Rinehart and Winston, 1968.

Ferber, Robert, *Statistical Techniques in Market Research*, New York, McGraw-Hill, 1949.

Festinger, Leon, *A Theory of Cognitive Dissonance*, New York, Row, Peterson and Company, 1957.

Gardner, Edward H., 'Consumer goods classification', *Journal of Marketing*, January 1945, pp. 275–276.

Grether, E.T., 'A theoretical approach to the study of marketing', in *Theory in Marketing*, Reavis Cox and Wroe Alderson, eds, Homewood, Irwin, 1950, pp. 113–123.

Harrigan, Katrun Rudie, *Strategies for Declining Businesses*, Lexington, Lexington Books, 1980.

Haspeslagh, Philipe, 'Portfolio planning: uses and limits', *Harvard Business Review*, January–February 1892, pp. 58–73.

Hedley, Barry, 'Strategy and the "Business Portfolio"', *Long-range Planning,* February 1977.

Hendrix, Philip E., *Product/Service Consumption: Implications and Opportunities for Marketing Strategy,* working paper, Emory University, 1986.

Holbrook, Morris B. and Howard, John A., 'Frequently purchased nondurable goods and services, in Robert Ferber, ed, *Selected Aspects of Consumer Behavior: A Summary from the Perspective of Different Disciplines,* Washington, DC, National Science Foundation, Directorate for Research Applications, Research Applied to National Needs, 1977, pp. 189–222.

Hollingworth, H.L., *Advertising and Selling,* New York, D. Appleton-Century, 1913.

Holloway, Robert J., 'An experiment on consumer dissonance', *Journal of Marketing,* January 1967, pp. 39–43.

Howard, John A., *Marketing Management: Analysis and Decision,* Homewood, Irwin, 1957.

Howard, John A. and Sheth, Jagdish N., *The Theory of Buyer Behavior,* New York, Wiley, 1969.

Huber, Joel, Holbrook, Morris B. and Kahn, Barbara E., 'Effects of competitive context and of additional information on price sensitivity', *Journal of Marketing Research,* August 1986, pp. 250–260.

Huff, David L., 'Defining and estimating a trading area', *Journal of Marketing,* July 1964, pp. 34–38.

Katona, George C., 'Rational behavior and economic behavior', *Psychological Review,* September 1953, pp. 307–318.

Katz, Elihu and Lazarsfeld, Paul F., *Personal Influence: The Part Played by People in the Flow of Mass Communications,* New York, Free Press, 1955.

Kauffman, Stuart., *At Home in the Universe: The Search for Laws of Self-Organization and Complexity,* London, Viking, 1995.

Kelly, Eugene and Lazer, William, eds, *Managerial Marketing: Perspectives and Viewpoints,* Homewood, Irwin, 1958.

Kleppner, Otto, *Advertising Procedure,* New York, Prentice-Hall, 1933.

Kotler, Philip, *Marketing Management Analysis, Planning and Control,* New York, Prentice-Hall, 1976.

Kuehn, Alfred A., 'Consumer brand choice as a learning process', *Journal of Advertising Research*, December 1962, pp. 10–17.

Lavidge, Robert J. and Steiner, Gary A., 'A model for predictive measurements of advertising effectiveness', *Journal of Marketing*, October 1961, pp. 59–62.

Levitt, Theodore, 'Marketing myopia', *Harvard Business Review*, July–August 1960, pp. 45–56.

Levitt, Theodore, 'Exploit the product life cycle', *Harvard Business Review*, November–December 1965, pp. 81–94.

Lewis, Richard J. and Erickson, Leo G., 'Marketing functions and marketing systems: a synthesis', *Journal of Marketing*, July 1969, pp. 10–14.

MacMillan, Ian C., 'Pre-emptive strategies', *Journal of Business Strategy*, Fall 1983, pp. 16–26.

Magee, John, 'The logistics of distribution', *Harvard Business Review*, July–August 1960, pp. 89–101.

Majaro, S., *International Marketing: A Strategic Approach to World Markets*, London, Allen & Unwin, 1985.

McCammon, Bert, 'Alternative explanations of institutional change and channel evolution', in Stephen A. Greyser, ed, *Toward Scientific Marketing*, Chicago, American Marketing Association, 1963, pp. 477–490.

McCammon, Bert, 'The emergence and growth of contractually integrated channels in the American economy', in Peter D. Bennett, ed, *Economic Growth, Competition, and World Markets*, Chicago, American Marketing Association, 1965, pp. 496–515.

McCarthy, E. Jerome, *Basic Marketing: A Managerial Approach*, Homewood, Irwin, 1960.

McGarry, Edmund D., 'Some functions of marketing reconsidered', in Reavis Cox and Wroe Alderson, eds, *Theory in Marketing*, Chicago, Irwin, 1950, pp. 263–279.

McKitterick, John B., 'What is the marketing management concept, in Frank Bass, ed, *The Frontiers of Marketing Thought and Action*, Chicago, American Marketing Association, 1957, pp. 71–82.

Mintzberg, Henry, 'The fall and rise of strategic planning, *Harvard Business Review*, January–February 1994, pp. 107–114.

Montgomery, David B. and Weinberg, Charles B., 'Toward strategic intelligence systems', *Journal of Marketing*, 43, 1979.

Nystrom, Paul H., *Retailing and Store Management*, New York, D. Appleton-Century, 1913.

Oxenfeldt, A.R., 'A multi-state approach to pricing', *Harvard Business Review*, July–August 1960, pp. 125–133.

Pederson, C.A. and Wright, M.D., *Salesmanship*, Homewood, Irwin, 1951.

Peters, Tom and Austin, Nancy, *A Passion for Excellence*, New York, Random House, 1985.

Petroshius, Susan M. and Monroe, Kent B., 'Effect of product-line pricing characteristics on product evaluations', *Journal of Consumer Research*, March 1987, pp. 511–519.

Porter, Michael E., 'How competitive forces shape strategy', *Harvard Business Review,* March–April 1979.

Porter, Michael E., *Competitive Strategy*, New York, Free Press, 1980.

Porter, Michael E., *Competitive Strategy: Techniques for Analyzing Industries and Competitors*, New York, Free Press, 1980.

Porter, Michael E., 'From competitive advantage to corporate strategy', *Harvard Business Review*, May–June 1987, pp. 43–59.

Prahalad, C.K. and Hamel, Gary, 'The core competencies of the corporation', *Harvard Business Review,* May–June 1990, pp. 79–91.

Reibstein, David J. and Gatignon, Hubert, 'Optimal product line pricing: the influence of elasticities and cross-elasticities', *Journal of Marketing Research*, August 1984, pp. 259–267.

Reilly, W.J., *Marketing Investigations*, New York, Ronald Press, 1929.

Revzan, David A., *Wholesaling in Marketing Organization*, New York, Wiley, 1961.

Roberts, E.B. and Berry, C.A., 'Entering new businesses: selecting strategies for success', *Sloan Management Review*, Spring 1985.

Robinson, Patrick J., Faris, Charles W. and Wind, Yoram, *Industrial Buying and Creative Marketing*, Boston, Allyn and Bacon, 1967.

Rogers, Everett M., *Diffusion of Innovations*, New York, Free Press, 1962.

Rogers, Everett M., 'New product adoption and diffusion', *Journal of Consumer Research*, 2, March 1976.

Russel, F.A., *The Textbook of Salesmanship*, New York, McGraw-Hill, 1924.

Schoeffler, Sidney, 'In defence of PIMS, GE and BCG', *Marketing News*, 9 February 1979.

Scott, W.D., *The Theory of Advertising*, Boston, Small, Maynard & Co., 1903.

Shaw, Arch W., *An Approach to Business Problems*, Cambridge, Harvard University Press, 1916.

Shaw, Arch W., 'Some problems in market distribution', *Quarterly Journal of Economics*, August 1912, pp. 706–765.

Sheth, Jagdish N., 'A theory of family buying decisions', in Jagdish N. Sheth, ed., *Models of Buyer Behavior: Conceptual, Quantitative, and Empirical*, New York, Harper & Row, 1974, pp. 17–33.

Sheth, Jagdish N., Gardner, David M. and Garrett, Dennis E., *Marketing Theory: Evolution and Evaluation*, New York, Wiley, 1988.

Slywotzky, Adrian J. and Morrison, David J., *The Profit Zone*, New York, Random House, 1997.

Smith, Wendell R., 'Product differentiation and market segmentation as alternative marketing strategies', *Journal of Marketing*, July 1956, pp. 3–8.

Tipper, H., Hollingworth, H.L., Hotchkiss, G.R. and Parsons F.A., *Advertising: Its Principles and Practices*, New York, Ronald Press, 1915.

Trout, Jack and Ries, Al, 'Positioning buts through chaos in market place', *Advertising*, May 1972.

Weld, L.D.H., *The Marketing of Farm Products*, New York, Macmillan, 1916.

Wenner, David L. and Leser, Richard W., 'Managing for shareholder value – from top to bottom', *Harvard Business Review*, November–December 1989, pp 52–68.

Wheeler, Ferdinand C., *The Technique of Marketing Research*, New York, McGraw-Hill, 1937.

White, Percival, *Market Analysis: Its Principles and Methods*, New York, McGraw-Hill, 1921.

White, Percival, *Sales Quotas*, New York, Harper & Bros, 1929.

White, Percival, *Marketing Research Technique*, New York, Harper & Bros, 1931.

Wiberg, O. and Albaum, G., *Selecting Export Markets*, working paper, Copenhagen Business School, 1986.

SOURCES ON THE DIGITAL SCHOOL OF MARKETING

The following sources are relevant for the digital school of marketing.

Arthur, W.B., 'Competing technologies, increasing returns, and lock-in by historical events', *Economic Journal*, 97, 1989.

Arthur, W.B., 'Positive feedbacks in the economy', *Scientific American*, 262, 1990, pp. 92–99.

Arthur, W.B., *Increasing Returns and Path Dependence in the Economy*, Ann Arbor, University of Michigan Press, 1994.

Arthur, W.B., 'Increasing returns and the new world of business', *Harvard Business Review*, 1996, pp. 100–109.

Besen, S.M. and Farrel, J., 'Choosing how to compete: strategies and tactics in standardization', *Journal of Economic Perspectives*, 8, 1994, pp. 117–131.

Biocca, Frank, 'Communication within virtual reality: creating a space for research', *Journal of Communication*, 42(2), 1992, pp. 5–22.

Blattberg, Robert C., Glazer, Rashi and Little, John D.C., eds, *The Marketing Information Revolution*, Boston, Harvard Business School Press, 1994.

Buchanan, J. and Stubblebine, W., 'Externality', *Economica*, November 1962, pp. 371–384.

Chatterjee, Patrali and Narasimhan, Anand, *The Web as a Distribution Channel*, Owen Doctoral seminar paper, 1994.

Chou, D. and Shy, O., 'Network effects without network externalities', *International Journal of Industrial Organization*, 8, 1990, pp. 259–270.

Cowen, Tyler., ed., *The Theory of Market Failure*, Fairfax, George Mason Press, 1988.

Coyle, Diane, *The Weightless World*, Cambridge, MIT Press, 1997.

Csikszentlmihalyi, Mihaly, *Flow: The Psychology of Optimal Experience*, New York, Harper and Row, 1990.

Cutler, B., The Fifth Medium, *American Demographics*, 12(6), 1990, pp. 24–29.

David, Paul. A., 'Clio and the economics of QWERTY', *American Economic Review*, May 1985, pp. 332–337.

Davis, Stan and Davidson, Bill, *2020 Vision: Transform Your Business Today to Succeed in Tomorrow's Economy*, New York, Simon & Schuster, 1991.

Davis, Stan and Meyer, Christopher, *Blur: The Speed of Change in the Connected Economy*, New York, Little Brown & Company, 1998.

Dennis, Everett E. and Pease, Edward C., 'Preface', *Media Studies Journal*, 8(1), 1994, pp. xi–xxiii.

Downes, Larry and Mui, Chunka, *Unleashing the Killer App*, Boston, Harvard Business School Press, 1998.

Drucker, Peter, *Management for the Future: The 1990s and Beyond*, New York, Plume, 1993.

Drucker, Peter, *Managing in Turbulent Times*, New York, Harper Business 1993.

Drucker, Peter, *Management Challenges for the 21st Century*, New York, Harper Business, 1999.

Durkheim, Emile, *The Division of Labor in Society*, translated by George Simpson, New York, Free Press, 1933.

Economides, Nicholas, 'A monopolist's incentive to invite competitors to enter in telecommunications services', in Gerard Pogorel, ed., *Global Telecommunications Strategies and Technological Changes*, Amsterdam, Elsevier, 1993, pp. 227–239.

Economides, Nicholas, 'Network externalities, complementarities, and invitations to enter', *European Journal of Political Economy*, 12, 1996, pp. 211–232.

Economides, Nicholas, 'Raising rivals' costs in complementary goods markets', *LECs Entering into Long Distance and Microsoft Bundling Internet Explorer*, discussion paper no. EC-98-03, Stern School of Business, New York University, revised March 1998.

Economides, Nicholas, 'The economics of networks', *International Journal of Industrial Organization*, 16(4), October 1996, pp. 673–699.

Economides, Nicholas and Himmelberg, Charles, 'Critical mass and network evolution in telecommunications', in Gerard Brock, ed., *Toward a Competitive Telecommunications Industry: Selected Papers from the 1994 Telecommunications Policy Research Conference*, Lawrence Erlbaum, 1995.

Economides, Nicholas and Himmelberg, Charles, 'Critical Mass and Network Size with Application to the US FAX Market', discussion paper no. EC-95-11, Stern School of Business, New York University, August 1995.

Economides, Nicholas and Woroch, Glenn, *Benefits and Pitfalls of Network Interconnection*, discussion paper no. EC-92–31, Stern School of Business, New York University, 1992.

Ellis, Howard S. and Fellner, William, 'External economies and diseconomies', *American Economic Review*, 33, 1943, pp. 493–511.

Farrell, J. and Saloner, G., 'Standardization, compatibility, and innovation', *Rand Journal*, 16, 1985, pp. 70–83.

Farrell, J. and Saloner, G., 'Installed base and compatibility: innovation, product preannouncements, and predation', *American Economic Review*, 76, 1986, pp. 940–955.

Fisher, Franklin M., *Disequilibrium Foundations of Equilibrium Economics*, Cambridge. Cambridge University Press, 1983.

Gilder, George, 'You ain't seen nothing yet', *Forbes*, 4 April 1988, pp. 88–93.

Gilder, George, *Microcosm: The Quantum Revolution in Economics and Technology*, New York, Simon and Schuster, 1989.

Gilder, George, 'Metcalf's law and legacy', *Forbes ASAP*, 13 September 1993, pp. 158–166.

Gilder, George, 'The bandwidth tidal wave', *Forbes ASAP*, 5 December 1994, p. 16.

Gilder, George, 'The coming software shift: telecoms', *Forbes ASAP*, 28 August 1995, pp. 147–162.

Gilder, George, 'Feasting on the giant peach', *Forbes ASAP*, 26 August 1996, p. 19.

Gilder, George, 'Fiber keeps its promise', *Forbes ASAP*, April 1997.

Glazer, Rashi, 'Marketing in an information intensive environment: strategic implications of knowledge as an asset', *Journal of Marketing*, 55, 1991, pp. 1–19.

Grossman, Lawrence K., 'Reflections on life along the electronic superhighway', *Media Studies Journal*, 8(1), 1994, pp. 27–39.

Grove, Andrew, *Only the Paranoid Survive*, New York, Doubleday, 1996.

Hafner, Katie and Lyon, Matthew, *When Wizards Stay Up Late: The Origins of the Internet*, New York, Simon & Schuster, 1996.

Hagel, John and Armstrong, Arthur, *Net Gain: Expanding Markets Through Virtual Communities*, Boston, Harvard Business School Press, 1997.

Hauser, John R., Urban, Glen L. and Weinberg, Bruce D., 'How consumers allocate their time when searching for information', *Journal of Marketing Research*, 30, 1993, pp. 452–466.

Hawkins, Donald T. 'Electronic advertising on online information systems', *Online*, 18(2), 1994, pp. 26–39.

Hayek, F.A., *Denationalization of Money*, 2nd ed., London, Institute of Economic Affairs, 1976.

Hock, Detlev J., Roeding, Cyriac R., Purkert, Gert, and Lindler, Sandro K., *Secrets of Software Success*, Boston, Harvard Business School Press, 1999.

Hoffman, Donna L. and Novak, Thomas P., 'Commercializing the information superhighway: are we in for a smooth ride?' *The Owen Manager,* 15(2), 1994, pp. 2–7.

Hoffman, Donna L. and Novak, Thomas P., *Marketing in Hypermedia Computer-Mediated Environments: Conceptual Foundations*, Project 2000 working paper no. 1., Owen Graduate School of Management, Vanderbilt University, 1995.

Hoffman, Donna L., Novak, Thomas P. and Chatterjee, Patrali, 'Commercial scenarios for the web: opportunities and challenges', *Journal of Computer-Mediated Communications* (special issue on electronic commerce), 1(3), 1995.

Kahn, Alfred E., *The Economics of Regulation: Principles and Institutions*, Cambridge, MIT Press, 1988.

Katz, Michael L. and Shapiro, C. 'Network externalities, competition, and compatibility', *American Economic Review*, June 1985, pp. 424–440.

Katz, Michael L. and Shapiro, C., 'Technology adoption in the presence of network externalities', *Journal of Political Economy*, September 1986, pp. 822–841.

Katz, Michael L. and Shapiro, C., 'Systems competition and network effects', *Journal of Economic Perspectives*, 8, 1994, pp. 93–115.

Katzner, Donald, 'Some notes on the role of history and the definition of hysteresis and related concepts in economic analysis', *Journal of Post Keynesian Economics*, 15(3), 1993, pp. 323–345.

Keller, Kevin Lane and Staelin, Richard, 'Effects of quality and quantity of information on decision effectiveness', *Journal of Consumer Research*, 14, 1987, pp. 200–213.

Keller, Kevin Lane and Staelin, Richard, 'Assessing biases in measuring decision effectiveness and information overload', *Journal of Consumer Research*, 15(4), 1989, pp. 504–508.

Kelly, Kevin, *New Rules for the New Economy*, New York, Penguin Group, 1998.

Kirkpatrick, David, 'A look inside Allen's think tank: this way to the i-way', *Fortune*, 11 July 1994, pp. 78–80.

Kitch, Edmund W., 'The nature and function of the patent system', *Journal of Law and Economics*, 20, 1977, pp. 265–290.

Klopfenstein, B.C., 'The diffusion of the VCR in the United States', in M.R. Levy ed., *The VCR Age*, Newbury Park, SAGE Publications, 1989.

Knight, Frank H., 'Some fallacies in the interpretation of social cost', *Quarterly Journal of Economics*, August 1924, pp. 582–606.

Kohli, Ajay K. and Jaworski, Bernard J., 'Market orientation: the construct, research propositions, and managerial implications', *Journal of Marketing*, 54, 1990, pp. 1–18.

Krugman, Paul and Obstfeld, Maurice, *International Economics: Theory and Policy*, New York, Addison-Wesley, 1997.

Lardner, James, *Fast Forward: Hollywood, the Japanese, and the onslaught of the VCR*, New York, W.W. Norton, 1987.

Lasswell, H.D., 'The structure and function of communication in society', in Lyman Bryson, ed., *The Communication of Ideas*, New York, Harper and Brothers, 1984.

Leibenstein, H., 'Bandwagon, snob, and veblen effects in the theory of consumer's demand', *Quarterly Journal of Economics*, 64, 1950, pp. 183–207.

Levinson, R.J. and Coleman, M.T., *Economic Analysis of Compatibility Standards: How Useful is It*, FTC working paper, 1992.

Licklider, J.C.R., 'Some reflections on early history', in Goldberg, Adele, ed., *A History of Personal Workstations*, New York, Addison-Wesley, 1988.

Liebowitz, S.J. and Margolis, S.E., 'The fable of the keys', *Journal of Law and Economics*, April 1990, pp. 1–26.

Liebowitz, S.J. and Margolis, S.E., 'Network externality: an uncommon tragedy', *Journal of Economic Perspectives*, 8, 1994, pp. 133–150.

Liebowitz, S.J. and Margolis, S.E., *Path Dependency, Lock-In, and History*, working paper, 1994.

Liebowitz, S.J. and Margolis, S.E., *Market Processes and the Selection of Standards*, working paper, 1994.

Liebowitz, S.J. and Margolis, S.E., 'Are network externalities a new source of market failure?', *Research In Law And Economics*, 17, 1995, pp. 1–22.

Liebowitz, S.J. and Margolis, S.E., 'Path dependence, lock-in and history', *Journal of Law, Economics, and Organization*, 11, 1995, pp. 205–226.

Liebowitz, S.J. and Margolis, S.E., 'Market processes and the selection of standards', *Harvard Journal of Law and Technology*, 9, 1996, pp. 283–318.

Martin, Chuck, 'Net future', New York, McGraw-Hill, 1999.

McKnight, L. and Bailey, J., eds, 'Special issue on internet economics', *Journal of Electronic Publishing*, Fall 1995.

McLuhan, Marshall, *Understanding Media*, New York, McGraw-Hill, 1964.

Meyer, Robert J. and Johnson, Eric J., 'Information overload and the non-robustness of linear models: a comment on Keller and Staelin', *Journal of Consumer Research*, 15, 1989, pp. 498–503.

Mills, Edwin, 'An aggregative model of resource allocation in a metropolitan area', *American Economic Review*, 57, 1967, pp. 197–210.

Modahl, Mary, *Now or Never*, New York, Harper Business, 2000.

Mohr, Jakki and Nevin, John R., 'Communication strategies in marketing channels: a theoretical perspective', *Journal of Marketing*, 54, 1990, pp. 36–51.

Moore, Geoffrey A., *Crossing the Chasm*, New York, Harper Business, 1991.

Moore, Geoffrey A., *The Gorilla Game*, New York, Harper Business, 1999.

Moore, Geoffrey A., *Inside the Tornado*, New York, Harper Business, 1995.

Moore, Gordon E., 'Some personal perspectives on research in the semiconductor industry', in Richard S. Rosenbloom and William J. Spencer, eds, *Engines of Innovation*, Boston, Harvard Business School Press, 1996, pp. 165–174.

Moore, Gordon E., 'Cramming more components onto integrated circuits', *Electronics*, 38(8), 1965, pp. 114–117.

Moore, Gordon E., *Lithography and the Future of Moore's Law*, paper presented to the Microlithography Symposium, 20 February 1995.

Moore, James F., *The Death of Competition: Leadership and Strategy in the Age of Business Ecosystems*, New York, Harper Business, 1996.

Moore, James F., 'Predators and prey: a new ecology of competition', *Harvard Business Review*, May–June 1993.

Movarec, Hans, *Robot*, Oxford, Oxford University Press, 1999.

National Academy of Sciences, *Realizing the Information Future: The Internet and Beyond*. 1994.

Peppers, Don, *The One to One Future: Building Relationships One Customer at a Time*, New York, Doubleday, 1997.

Drucker, Peter F., *Post Capitalist Society*, New York, Harper Collins, 1993.

Keen, Peter G.W., *Competing in Time: Using Telecommunications for Competitive Advantage*, New York, Ballinger Publishing Company, 1988.

Press, Larry, 'The internet and interactive television', *Communications of the ACM*, 36(12), 1993.

Reid, Elizabeth M., *Electropolis: Communication and Community on Internet Relay Chat*, honours thesis, Department of History, University of Melbourne, 1991.

Harris, Robert D., *The New Paradigm of Business: Emerging Strategies for Leadership and Organizational Change*, Pedigree Books, 1993.

Rose, Chris, 'Burn those bank notes – digital cash is coming', *Power PC News*, 1(10), document no. 3032, 1994.

Saffo, Paul, *Disinter-Remediation: The Surprising Impact of Information Systems on Markets and Organizations*, Business & Technology In a Digital Economy Conference, October 1997.

Schelling, Thomas C., *Micromotives and Macro Behavior*, New York, Norton. 1978.

Schickele, Sandra, *The Internet and the Market System: Externalities, Marginal Cost, and the Public Interest*, Proceedings from the International Networking Conference, 1993.

Schwartz, Evan I., 'Fran-on-demand', *Wired*, September 1994, pp. 60–62.

Seybold, Particia, *Customers.com*, New York, Random House, 1998.

Shapiro, Carl and Varian, Hal R., *Information Rules*, Boston, Harvard Business School Press, 1999.

Smith, Adam, *The Wealth of Nations*, London, W. Strahan & T. Cadell, 1776.

Steuer, Jonathan, 'Defining virtual reality: dimensions determining telepresence', *Journal of Communication*, 42(4), 1992, pp. 73–93.

Stewart, David W. and Ward, Scott, 'Media effects on advertising', in Jennings Bryand and Dolf Zillman, eds, *Media Effects, Advances in Theory and Research*, Hillsdale, Lawrence Erlbaum Associates, 1994.

Tapscott, Don, *Blueprint to the Digital Economy*, New York, Mcgraw-Hill, 1998.

US Congress, Office of Technology Assessment, *Electronic Enterprises: Looking to the Future, OTA-TCT-600*, Washington, DC, US Government Printing Office, May 1994.

Urban, Glen, Weinberg, Bruce and Hauser, John R., *Premarket Forecasting of Really New Products*, working paper, Massachusetts Institute of Technology, 1994.

Van Raaij, W. Fred, 'Postmodern consumption', *Journal of Economic Psychology*, 14, 1993, pp. 541–563.

Venkatesh, Alladi, Sherry, John F. Jr. and Firat, A. Fuat, 'Postmodernism and the marketing imaginary', *International Journal of Research in Marketing*, 10, 1993, pp. 215–223.

INDEX
....